PUBLIC HEALTH IN THE 21ST CENTURY

HYGIENICALLY RELEVANT BIOFILMS

PUBLIC HEALTH IN THE 21ST CENTURY

Additional books in this series can be found on Nova's website
under the Series tab.

Additional e-books in this series can be found on Nova's website
under the e-book tab.

PUBLIC HEALTH IN THE 21ST CENTURY

HYGIENICALLY RELEVANT BIOFILMS

ROK FINK

nova publishers
New York

Library of Congress Cataloging-in-Publication Data

ISBN: 978-1-63463-316-1
Library of Congress Control Number: 2014955174

Published by Nova Science Publishers, Inc. † New York

CONTENTS

PREFACE

This book describes biofilms in the context of hygiene in food safety and hospital facilities, starting with bacterial adhesion as an initial step of cell colonization, followed by the growth and maturation of biofilms. The main focus of this book is on biofilm prevention and control. The resistance of bacteria to several antibiotics is a global public health problem, not limited to hospital facilities, but also to other sectors, including the food industry. This raises questions about whether effective cleaning and disinfection reduce the risk of food-borne and hospital-acquired illness. Can the surface characteristics be treated to prevent bacterial adhesion and, consequently, the formation of biofilms? Biofilms are complex communities of bacteria attached the surfaces, with fully developed architecture and highly organized interactions. A better understanding of the biofilm lifecycle, with advanced strategies of physical, chemical and biological actions, would reduce the risk of infection.

The question is not whether the preventive and control measures will play a significant role in public health, but rather what the implications of their use will be for hygiene in the food industry and hospital facilities.

Assist. Prof. Dr. Rok Fink

ACKNOWLEDGMENTS

The author would like to express his sincere appreciation to Assoc. Prof. Borut Poljšak, Ph.D., Assist. Prof. Mojca Jevšnika, Ph.D., and Sebastjan Filip, Ph.D., for their valuable comments and suggestions when reviewing the manuscript. Thanks Terry Troy Jackson for grammar proof.

At last but not the least I want to thank my closest ones who appreciated me for my work and motivated me in writing this book.

Chapter 1

INTRODUCTION TO HYGIENICALLY RELEVANT BIOFILMS

"It's hard to beat someone who never gives up"
(Babe Ruth)

Since the mid-20th century, bacterial infections have been recognized as a significant public health issue. [1] The U.S. National Institutes of Health reports that more than 80% of bacterial infections are related to biofilms; moreover, the Center for Disease Control has declared biofilms to be the most important clinical problem. [2, 3] Several foodborne illnesses are related to biofilm formation on food contact materials e.g. *L. monocytogenes*, *E. coli*, *S. aureus*, etc. [4-6] Furthermore, hospital-acquired infections often cause illnesses such as lung infection, cystic fibrosis, colitis, urethritis, and endocarditis. [7] All these facts suggest that biofilm hygiene should become a public health priority, especially for hygiene decision makers.

BIOFILM DEFINITION

A biofilm is a community of microbial cells that is irreversibly attached to a surface and enclosed in a polysaccharide matrix. In addition to cells in the biofilm matrix, non-cellular materials, i.e. organic and inorganic matter, are also present. Organisms in biofilms also differ from planktonic cells with respect to gene transcriptions and interactions. Microorganisms can occupy a wide variety of surfaces, including living tissues, internal medical devices, wood, stainless steel, industrial or potable water system piping, and natural aquatic systems. [8, 9]

In natural environments, microorganisms can persist as planktonic organisms, which are individual cells freely suspended in a liquid medium, or as a complex community as biofilms. In recent years, the definition of a biofilm has been constantly changing, since each new study increases the existing knowledge about biofilm formation, structure, maturation, and resistance. Based on the explorations of their known characteristics, and their newly discovered physiological properties, biofilms are defined as "structural communities of microorganisms that are irreversibly attached to a surface or interface, embedded in an extracellular matrix of polymeric substances which these cells have produced". [10]

Biofilm growth is a complex process that involves several steps (Figure 1). First, it is necessary that the contact surface become preconditioned, either intentionally or by the adhesion of macromolecules present in a bulk fluid. These favorable conditions force the microorganisms to adsorb to the surfaces, in either a reversible or irreversible manner. The persistent microorganisms that remain on the surface start to produce signaling molecules and secrete extracellular polymeric substances (EPS) (Figure 1). [11] EPS are a complex network formed essentially by polysaccharides, proteins, phospholipids, teichoic acids and by nucleic acids. EPS can also contain crystals, silt particles, milk residues [10] and even blood components or other organic and inorganic components due to the conditions under which the biofilms were formed. [6, 12]

Biofilm growth and development are determined by the convective and diffusional transport of oxygen and nutrients through channels. Biofilm formation also includes proliferation and cell growth, accompanied by the secretion of EPS and the transport of waste products to the exterior of the biofilm. [10] Cells and nutrients are simultaneously transported and accumulated on the surface, and some parts of the biofilm are removed by the surrounding streams, providing a balance of growth and detachment events. [13] This ecosystem allows bacterial cells to survive extremes in environment, such as those during sanitizing and disinfection. In biofilms, planktonic organisms assume a temporary multicellular organization, in which "group behavior" enable survival in host environments.

Figure 1. Growth stages of biofilm from adhesion to EPS production and maturation.

Biofilms can occupy almost any materials that have application in the food industry and medicine. [14] In order to understand biofilms and principles for controlling them, it is vital to understand bacterial adhesion as an initial phase of biofilm growth. Biofilm formation, growth, maturation and dispersion are affected by several factors that can determine the survival or death of biofilms. Knowing the conditions under which biofilms form enables the prevention and control of them to be performed more efficiently. [10] What was in past defined as the formation of a community of microorganisms attached to a surface has become a complex developmental process that is multifaceted and dynamic in nature. The transition from planktonic growth to biofilm occurs in response to environmental conditions, and involves multiple regulatory networks, which translate signals to concerted gene expression and cause the spatial and temporal reorganization of the bacterial cells. [15-20] This cellular change alters the expression of surface molecules, nutrient consumption, and virulence and assures bacteria the properties that enable their survival in unfavorable conditions. [21-23]

Evolution of Biofilm Science

The first observation of biofilms is considered to have been provided by Antonie van Leeuwenhoek in 1684. He noticed that microbes attach to the tooth surface, forming kind of "film." In 1923, Angst reported that marine bacteria on the surfaces of ship hulls caused serious corrosion damage. By the 1980s, biofilms had been found on surfaces of several materials including industrial material, medical devices, tooth decay, vein and urinary catheters, contact lenses, food contact materials, pipe systems, etc. [24]

Heukelekian and Heller [25] observed the so-called "bottle effect" for microorganisms. They found that bacterial growth and activity were substantially enhanced by the incorporation of a surface onto which these organisms could attach. Zobell [26] reported that the number of bacteria on surfaces was significantly higher than in the surrounding medium. However, it was not until the introduction of the electron microscope, allowing high-resolution photo-microscopy at much higher magnifications than light microscopes, that a detailed examination of biofilms was possible. Jones et al. [27] used scanning and transmission electron microscopy to examine biofilms on filters in a wastewater treatment plant and found that biofilms are composed of a variety of organisms. Using a specific polysaccharide stain called ruthenium red, and combining this with an osmium tetroxide fixative, early researchers were also able to determine that the matrix material surrounding and enclosing cells in these biofilms was polysaccharide. As early as 1973, Characklis [28] studied microbial communities in industrial water systems and showed that they were not only quite persistent but also highly resistant to disinfectants, such as chlorine. Based on observations of dental plaque and sessile communities in mountain streams, Costerton et al. [29] in 1978 explained the mechanisms through which microorganisms adhere to living and non-living materials and the benefits of this ecologic niche.

The term "biofilm" had been used sporadically among researchers for some years before it was considered to be a term acceptable for use in publications. The earliest use of "biofilm" can be found in the publication "Half-order reactions in biofilm and filter kinetics," published in 1977.

Since then, studies of biofilms have been becoming increasingly relevant for public health, taking into account material science, physics, microbiology, biochemistry and other fields. Much of the work since the 1990s has relied on methods such as scanning electron microscopy (SEM) or standard microbiologic culture techniques for biofilm determination. Two major inventions since the year 2000 have dramatically changed the understanding of biofilms: the utilization of the confocal laser scanning microscope to characterize biofilm ultrastructure, and an investigation of the genes involved in cell adhesion and biofilm formation. [8] In recent years, mathematical modelling has been able to predict the growth rate of different biofilms under various conditions, allowing analysis of biofilms without growing them.

WHY DO BACTERIA FORM BIOFILMS?

Survival is a biological principle known to all living beings, and although different organisms have developed their own survival strategies, all have one common goal that relates survival with the ability to adapt to dynamic changes in the environment. Microorganisms are particularly adaptable to changes in the environment because of their high reproduction rates, which allows them to transfer genes to future generations in short periods of time. [30]

In addition to bacterial characteristics, surfaces also seem to play an important role in the survival and dissemination of microbial cells. In fact, there are factors that contribute to the adhesion of microorganisms near the surface: the concentration of nutrients, pH, oxygen, and the hydrodynamic force that is exerted on the adhered cells. The adhesion of cells to surfaces represents an advantage and is a prevalent form of survival in hostile environments. [31] Other important factors of biofilm formation are the genetic load and regulation of microorganisms, as well as the properties of the adhesion surface, the type of bacterial cells, hydrodynamic conditions and other environmental factors. [6]

Attachment is a complex process regulated by diverse characteristics of the growth medium, substratum, and cell surface. An established biofilm structure comprises microbial cells and EPS, has a defined architecture, and provides an optimal environment for the exchange of genetic material between cells. Another advantage of biofilms is that cells can also communicate via quorum sensing (QS), which may in turn affect biofilm processes, such as detachment.

WHY ARE BIOFILMS A HYGIENE PROBLEM?

Biofilms are extremely difficult to eradicate, mainly due to several adaptation strategies; therefore, the prevention of biofilm formation would be the optimal strategy for control. Although biofilm removal can occur naturally by intrinsic processes, eradication by human action is considered to be very expensive because the use of sanitizers, disinfectants and mechanical force is time consuming.

Currently, there is no proper technique or measure to control or prevent biofilm formation without undesirable side effects. [10] Biofilms have a wide diversity of defense mechanisms. In comparison to planktonic cells, biofilm cells are 10 to 1000 times more resistant to antimicrobial agents. [32] Some authors reports that the function of EPS as a polymer is its interaction with antimicrobials, decreasing their activity, before they can reach the cells in the matrix. [33-36] A more profound understanding of biofilm resistance is necessary in order to increase the effectiveness of biofilm control strategies. Pathogens such as *Listeria*, *Pseudomonas*, *Bacillus* and *Salmonella* can form biofilms in food contact materials, thus representing a potential threat to public health from outbreaks of food-borne diseases. [37] In addition, biofilm formation in hospital environments, e.g. implants, catheters and vascular stents, can cause chronic inflammation. Especially considering the fact that patients have a weakened immune system, any contact with biofilms can have life-threating consequences. Common to both types of biofilms is thriving under physiological conditions, making them more difficult to eliminate.

A paradigm of good hygiene practice is that the biofilms are resistant to antimicrobial agents, which makes their elimination significantly challenging (Figure 2). Good hygiene must balance cleaning procedures and biofilm resistance. Use of proper sanitizers and cleaning methods will prevent EPS production and inhibit QS. Therefore, the selection of proper control under sufficient conditions will determine whether the biofilm will survive on a surface or not (Figure 2).

Bacteria in biofilms resist antibiotics via several mechanisms, including:

- decreased penetration or diffusion of antimicrobial agents into biofilms,
- increased activity of multidrug efflux pumps,
- QS systems,
- starvation or stress responses,
- gene transfer that turns susceptible planktonic cells into antibiotic-resistant ones. [38]

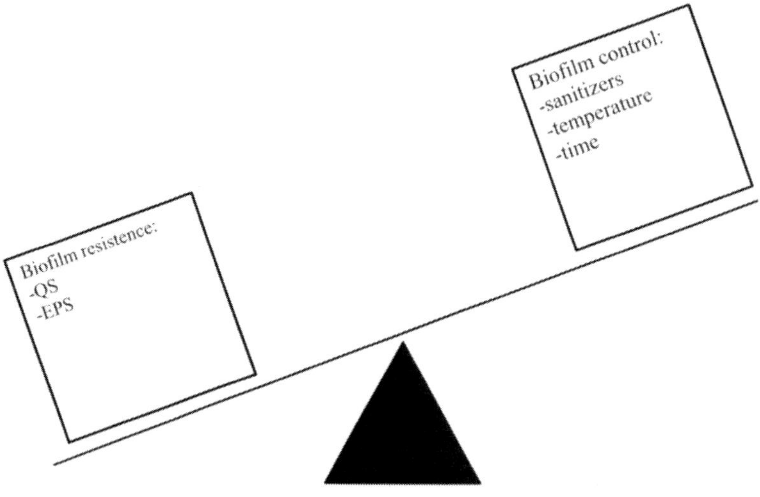

Figure 2. Paradigm of good hygiene practice related to biofilm control.

Another hygiene challenge is multispecies biofilms; monoculture biofilms are rarely present in nature. [39] Burmølle et al. reported that mixed-species biofilms have shown less susceptibility to sanitizers and disinfectants in comparison to single species biofilms. One reason for this can be found in their complex architectures and the interactions between organisms. [40] Some studies report that the use of sanitizers may lead to the detachment of biofilms when bacteria are not inactivated, so they could reattach and grow again. [41-43] However, when biofilms are inactivated, but not removed, they may provide a good foundation for the attachment of living bacteria. [44, 45] Interactions in multispecies biofilms affect the function and physiology properties, resulting in enhanced resistance and virulence, which is of substantial importance to human health and activities. The underlying mechanisms causing these synergistic effects can be determined at the molecular and evolutionary levels, and further exploration is now possible due to the enhanced research and higher throughput of available techniques. [40]

Biofilms have great importance for public health because of their role in certain food-borne illnesses and their significance in a variety of hospital-acquired infections. A greater understanding of biofilm processes should lead to innovative and effective strategies for biofilm control and a resulting improvement in patient management.

Food Industry

Pathogens and spoilage organisms on food and food contact materials causing the formation of biofilms remain an extremely serious problem in the food industry. Møretrø and Langsrud [46] reported that outbreaks associated with *L. monocytogenes* are one of the five major organisms causing foodborne diseases, and account for approximately one third of the deaths resulting from foodborne illnesses. Researchers have linked these outbreaks to the inability to eradicate the organisms from contaminated food contact materials and environments. Some risk areas have been identified in food operations that promote biofilm formation, including slicing, packaging, cutting, dosing, skinning, pumping and mixing equipment. [6] The resistance of pathogens is based on the fact that organisms have formed a highly structured community of cells that have shown great ability to adapt to sanitizing and disinfection procedures. [47] Several studies in the field of biofilms in the food industry that compared planktonic cells to mature biofilms reported increased resistance to most growth-limiting factors, including nutrient deficiency, low pH, low water activity, and the presence of salts, antibiotics, and antimicrobials. [35, 48, 49] Pathogen contamination problems and biofilm development are also the results of food residues that remain after inadequate cleaning, in which case bacteria can grow more rapidly, and the expression of the biofilm phenotype may not be required. Removal of food residues is often more easily mastered than the removal of mature biofilms. Therefore, it is becoming increasingly critical to gain a better understanding of the physicochemical, biochemical and genetic factors that govern the growth and persistence of biofilms. [50]

Hospital Facilities

Biofilms in hospital environments represent a major public health problem, since artificial surfaces in contact with biological systems are substrates for growth, thus accounting for serious life-threatening complications, the consequences of which expensive for healthcare systems. Infections caused by biofilms are difficult to eradicate, because the cells can survive in the physiological environment. [36, 51-53] Nowadays, many people have internal devices, often as permanent implants. The most important failures are hygiene failures, when microbial colonization develops on biomaterial surfaces. [54-56] Hospital-acquired infections cause two million deaths annually in the United States and lead to more than $5 billion in added medical costs per annum. [57]

The resistance of biofilms to antimicrobial treatment is a global healthcare issue. The number of bacteria species that are resistant to antibiotics is a fact that must be understood by decision makers, health worker, patients and their relatives. In the past, resistant bacteria were limited only to intensive-care units, but increasing numbers of cases are currently present in extended-care facilities, ambulatory surgical units, dentists, home healthcare sites, and other healthcare settings. For example, *S. epidermidis* is now recognized as common organism causing hospital-acquired infections. [58]

In addition to the hygiene and financial burdens they impose, biofilms have an impact on the energy requirements of the system, cause corrosion, and increase frictional drag.

PREVENTION AND CONTROL OF BIOFILMS

The persistence of microbial biofilms represents a significant challenge to the establishment and maintenance of hygienic conditions in various environments. The use of antibacterial agents is vital to the hygienic control of the microbial population. Such agents act against planktonic cells but are inappropriate to control adherent biofilm communities. The most important prevention measure against such formations is regular sanitizing and disinfecting, thus preventing bacteria from forming EPS. Cleaning in the food industry and hospital environment should be done in a manner that can break up or dissolve the EPS so that disinfectants can have access the bacterial cells. Recent findings promote the use of bacteriophages, natural antimicrobials, combinations of disinfectants, bio-sanitizers and other novel strategies to combat biofilms. [59-61]

The resistance of biofilm communities to antibacterial agents has been attributed to the organization of cells within an EPS. This fact, in combination with the proximity of cells, causes limitation of the access of an agent from its point of application to the deeper-located cells. Since these cells are protected, they survive and will multiply and divide rapidly when the environment is less hostile, e.g. if the bulk of the treatment agent is depleted or the exposure is transient. One of the features of biofilms is that nutrients become depleted within the core of the developing community. This causes the establishment of the spatial gradients of growth rate and redox within the community structure. Absorption of nutrients is not equal across the entire biofilm surface. Faster-growing and more susceptible cells will generally lie on the periphery of the biofilm with slow-growing cells being more deeply placed. [62, 63]

The more resistant bacterial cells will survive exposure to antibacterial agents. Repeated exposure to sub-lethal treatments of antibacterial agents might not only cause resistance within the population, but the bacteria can also share this resistance with third-party agents. Since neither mechanism can provide a detailed explanation of resistance, together they will delay the eradication of the exposed microbial population and allow other selection and regulation events to occur. [64]

Surfaces are cleaned to remove dirt and some portion of microorganisms and are then disinfected to inactivate any microorganisms that were not eliminated by sanitizing. The purpose of cleaning and disinfecting is to reduce the amount of microbial cells so that the microbiological quality is satisfactory. It must be realized that on the surfaces of food contact materials some bacterial cells always survive sanitizing and disinfection processes. For evaluating cleaning and disinfection, efficacy swabbing techniques used to sample surfaces do not detach all the microbial cells, and the removed portion is lower after cleaning and disinfection. The second concern is that although some cells are not culturable, they may still be alive. The removal of biofilms is achieved via a combination of four factors: formulations and concentrations of cleaning and sanitizing agents, exposure time, temperature, and mechanical activity.

Antimicrobial agents are active substances or preparations that contain one or more active substances, which are presented to the user in their final form, and whose function is to destroy, stop the growth, make harmless, avoid or control by any mean the action of a pathogenic organism by a biological or chemical process. Although agents are aimed at the reduction of microorganisms, their use does not necessarily reduce biofilm formation. It is very important to use the correct quantity of the proper agent, with the correct frequency. Incorrect application is expensive and leads to adverse effects. [65] Current methods for microbial population control include the application of a wide variety of chemical compounds, such as alcohols, aldehydes, bisphenols, diamidines, halogen releasing agents, peracetic acid, heavy metal derivatives, peroxygens, phenols, quaternary ammonium compounds, chlorine-based agents and ozone. [66]

Bacterial strains react individually to each antimicrobial agent, either because of their characteristics or due to resistance mechanisms. Therefore, when selecting antimicrobial agents, it is important that an efficacy evaluation regarding the eradication of microbial cells present on the surface be performed. Only after acquiring information about the characteristics of the bacterial population to be eliminated is it possible to define the relation between the minimum inhibitory concentration and the contact period of an antimicrobial agent. [67] Ferreira et al. [68] reported four major categories of antimicrobial agents with regard to their efficacy: the oxidants, the electrophilic agents, the cationic membrane biocides, and the weak acids. Oxidants act via radical-mediated reactions of oxidizing organic material; electrophilic agents react covalently with cellular nucleophiles to inactivate enzymes; cationic membrane active biocides destabilize membranes leading to rapid cell lysis; the weak acids interfere with the ability of the cell membrane to maintain a proper pH balance, resulting in the acidification of the cell interior and widespread metabolic disruption. [68]

In addition to chemical agents, the removal of a mature biofilm will also require extensive mechanical action, such as scrubbing or scraping. Sanitizers flowing over the surface remove only the top layer and exposes the subsequent layers to nutrients, which results in increased biofilm growth. As bacterial cells are released from the EPS matrix of a biofilm, they are much less resistant to subsequent antimicrobial agents used. Although the EPS of the biofilm will affect the efficiency of removal, there are several combinations of sanitization and disinfection agents that have been successfully implemented. In general, the greater the contact time is, the more effective the antibacterial agent is. Moreover, any chemical treatment combined with mechanical forces will increase the efficiency of the removal of biofilms. Therefore, circulating water for clean-out-of-place (COP) tanks, pipelines, floor scrubbers, or brushes or scrapers is highly recommended. Furthermore, care should be taken when working with force since some brushes and scrapers may be abrasive and leave scratches on surfaces, further increasing the load of dirt and consequently biofilm formation. Instructions for cleaning have to be precise and detailed in order to maximize the potential of an antibacterial agent. [69]

Gibson et al. recommends the following protocol: First, add parts (0.5 oz./gal, pH 11-12) to a chlorinated alkaline detergent in circulating water at 70 °C for 20 minutes. Second, rinse the parts with water and place them in a circulating 70 °C water with phosphoric acid (1 oz./gal) for 20 minutes. Then, rinse with potable water and place the parts in a chlorine solution (0.3 oz./gal) for 15 minutes. Finally, rinse with potable water. This protocol is effective for combating biofilms and has been found to be extremely effective on biofilms containing organisms such as *S. aureus*. [70] Since the composition of the EPS varies among the species, a combination of sanitizers and disinfectants may be required to eliminate the biofilm from the surface. Some ions, especially cations, play a role in bacterial adhesion. Their absence often results in the detachment of bacteria. Therefore, chelators in sanitizers may be effective in the removal of biofilm. [70]

CONCLUSION

Research in biofilms has increased exponentially in the past decade. The biofilm phenomenon is related in large part to its relevance to health. The capability of microorganisms to adapt to their surrounding hostile environment is remarkable, but whether the cells actually do communicate to each other, coordinate the formation of biofilms, and specialize their roles within biofilms for the benefit of the community remains debatable. [71] Biofilms in the food industry and hospital facilities are distinct types, although both cause infection in the physiological environment. Biofilms on food and food contact material can be consumed, causing food-borne infections and intoxication. Meanwhile, biofilms on medical devices are in direct contact with patients and can generate hospital-acquired infections. These are some of the reasons hygiene is vital in the prevention and control of biofilms.

Chapter 2

MECHANISMS OF BACTERIAL ADHESION TO SURFACES

"Irreversibility is the mechanism that brings order out of chaos"
(Ilya Prigogine)

Interactions between bacteria and contact materials play an important role in the technology of food production, as well in hospital environments. [72] Bacterial adhesion is the first step in colonization, followed by the proliferation and formation of biofilms. [10]

Bacterial attachment to the surface materials or other cells is a physiochemical phenomenon defined by the Van der Waals and electrostatic forces acting between the cells and the attachment surface. [4, 73, 74] Among microorganisms, bacteria are the main compounds of biofilms and are about 0.5 to 2 μm in size, which is nearly in the range of colloidal particles. Therefore, interactions between the attachment surface and bacteria can be explained by the theory of colloidal particles.

DERJAGUIN, LANDAU, VERWEY AND OVERBEEK'S (DLVO) THEORY

The DLVO theory has been used by several authors to define bacterial adhesion to surfaces. [75-79] The interaction between the surfaces and bacteria can be explained as the sum of van der Walls and Coulomb interactions. Since the van der Walls attractive force is stronger close to the surface, cells cannot separate from the surface merely by Brownian motion and are, therefore, attached irreversibly. In contrast, Coulomb interactions become dominant at a distance away from the surface because the van der Waals force decreases with distance (Figure 3). [80]

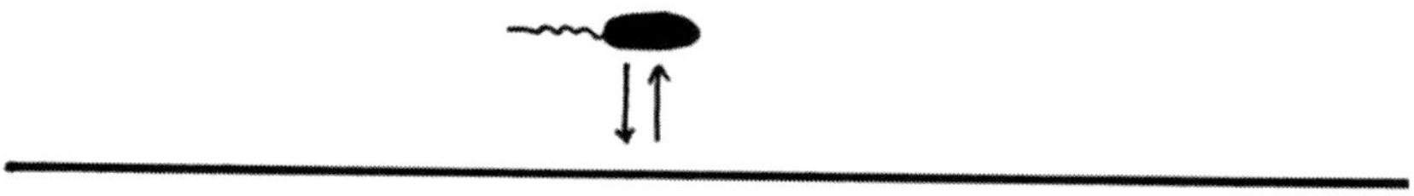

Figure 3. Interactions between bacteria and surface Van der Walls and Coulomb forces.

If in the aqueous solution charge particles are present, counter ions against the surface charge are attracted by the particle, forming an electric double layer. Since bacteria and other natural surfaces in an aqueous solution are usually negatively charged, repulsive electrostatic energy is caused by the overlap of the electrical double layers of bacterial cells and the surface. This repulsive energy increases as the ionic strength of an aqueous solution decreases because the shielding of the surface is charged by the ions in the electrical double layers. At low ionic strengths (Figure 4a), when bacterial cells move close to the surface, there is an energy barrier, which the cells cannot overcome by motility or Brownian motion. In such cases, there is a small secondary energy minimum outside of the energy barrier. Meanwhile, in increased ionic strengths (Figure 4b), the electric double layer shrinks, which allows the cells to become closer. The distance from the surface to the secondary energy minimum is usually several nanometers, depending on the ionic strength of media. [81]

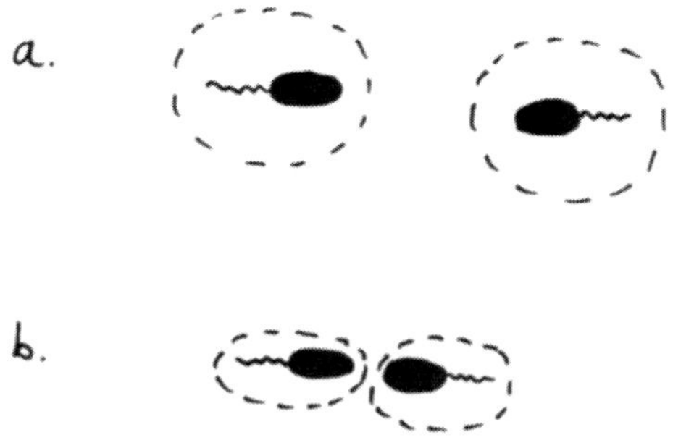

Figure 4. Impact of ionic strength on bacterial adhesion.

In the initial step of cell adhesion, a bacterial cell comes close to the surface by its motility or Brownian motion, and adheres to the surface, reversibly. In the next step, the bacterial cell uses appendices, such as pili and flagella, or starts to secrete EPS, which can overcome the energy barrier due to their small radii, for bridging between the cell and surface. [10] When the energy barrier becomes higher and farther from the substratum at lower ionic strengths, it becomes difficult for the appendices and EPS to interact with the substratum, and bacterial cells start to detach. In other cases, when high ionic strengths are present, the energy barrier is overcome and bacterial cells can easily and rapidly attach irreversibly to the surface. [82]

Some experimental studies have shown that the DLVO theory is important for understanding bacterial adhesion, but have also revealed some differences for bacterial strains between experiments and theory. [81] Poortinga et al. [81] showed that DLVO has the same discrepancies for *Pseudomonas*, *E.coli*, *Rhodococcus* and *Corynebacterium* when comparing theory and experiments.

The DLVO theory can explain only the interactions between long-range particles, thus allowing only the prediction that bacterial cells approach up to a certain distance from the contact surface. Nevertheless, the theory can only partially explain of the adhesion phenomena. Therefore, Van Oss et al. [83] proposed an extended DLVO theory that also includes the thermodynamic aspects of adhesion. This theory considers three fundamental interactions: Lifshitz-van der Waals, electrostatic and Lewis acid-base forces. The total energy of these interactions is the result of the sum of the free energies of each interaction force.

Bacterial Cell Surface Characteristics

Bacterial cell characteristics play an important role in the process of adhesion. In particular, characteristics such as surface charge, cell wall hydrophobicity, cell surface structures, as well as the type of excreted EPS are important for the adhesion process. Different bacterial species and strains can adhere to different material surfaces; this is due to the variability of physical and chemical characteristics between bacterial species and their strains. [84]

Cell Surface Structure

The differences in the cell surface structure are related to the distinctions in cell wall architecture and the presence of biomolecules found on the cell wall. Not only the interaction between the surfaces, but also outer cell surfaces structures and molecules that attached on the cell surface are important in bacterial adhesion. These structures determine the physical and chemical surface properties of bacterial cells and, therefore, determine adhesion and biofilm formation properties. [4]

As previously stated, the EPS on the cell surface play a crucial role in initial cell adhesion. Long et al. [85] removed the EPS from the cell walls of different strains of bacteria and studied the deposition on silica particles at several ionic strengths. The electro-potential and the size of the bacteria were not changed in comparison to the untreated bacteria. Nevertheless, the authors determined that the deposition rates for the untreated bacteria were consistently higher than for the treated ones, concluding that the absence of EPS decreases bacteria cell deposition on surfaces. [85]

Some researchers have shown that genetically specific interactions can be triggered by the surface chemistry and the fluid conditions. Floka et al. [86] showed that using *S. epidermidis* combinations of interactions between the bacterial cells and the surface including the chemical functionality of the surface and the presence of shear stress significantly affected the expression of genes in the regulation of biofilm formation, which in turn regulated the production of a key EPS. [86]

Studies also show that the presence of inorganic phosphates has significant influence in the biofilm formation of *P. fluorescens* and, depending on the concentration found in the environment, determined the adhesive action of LapA, which is an adhesin located outside the bacterial cell membrane. [87, 88] Another study proposed that in the presence of low concentrations of inorganic phosphate, cell detachment is induced through several internal molecular mechanisms leading up to the autolytic action on LapA. They also found that it promoted cell detachment and the return to a planktonic mode of *P. fluorescens*. Vrouwenvelder et al. [89] analyzed how low levels of phosphate affect the biofouling of reverse osmosis membranes, and proved that biofouling at low phosphate concentrations decreases the biomass growth. [89]

In addition to the electro-kinetics of bacterial cell wall, its architecture also plays a crucial role in the process of biofilm growth. The main differences are between Gram-positive and-Gram negative bacteria. [90] The bacterial cell wall of Gram-positive cells is constructed of a relatively thick peptidoglycan layer, consisting of cross-linking carbohydrates and peptides, which acts as an intensive and flexible barrier capable of resisting significant levels of external stress. The Gram-positive cells usually possess appendages covalently attached to the peptidoglycan layer or the inner plasma membrane. [91] In contrast to Gram positives, the cell walls of Gram-negative bacteria consist of a thinner peptidoglycan layer, which has an outer membrane consisting of proteins, lipopolysaccharides and phospholipids. [92] Differences among Gram-positive and Gram-negative bacteria can be found in the ratios of the various cell wall appendages or their chemical compositions. These are also the factors defining the variations in the properties of different species and even different strains of the same species. [74] These different structures explain the variety in the chemical properties of the bacterial surfaces. Bacterial surfaces with low ratios between nitrogen and carbon are more hydrophobic, due to the presence of proteins. In contrast, hydrophilic surfaces have a high oxygen and carbon ratio. [93, 94] Hydrophobicity or hydrophilicity depends on the strain and on the stage of maturity of the microorganisms or the bulk environment. [80]

Bacterial adhesion is influenced by surface cell hydrophobicity and material properties. Two of the most often studied chemical properties of bacteria cell are hydrophobicity and surface charge, both of which depend on the architecture of the cell external structure. [95, 96]

Cell Hydrophobicity

The hydrophobic interaction is an entropic effect originating from the disruption of highly dynamic hydrogen bonds between molecules of liquid water by the nonpolar solute. Bacterial adhesion in aqueous solutions usually involves cell wall hydrophobic groups, especially nonpolar groups' surface proteins, which allows cells to approach the surface, followed by changes in surface polymers leading other functional groups to approach the surface for the formation of short-range attractive polymeric interactions. [97]

It is generally known that bacteria with hydrophobic properties prefer hydrophobic surfaces and *vice versa* hydrophilic bacteria prefer hydrophilic surfaces. Furthermore, hydrophobic bacteria tend to have greater adhesion than hydrophilic bacteria. [98] Pang et al. showed that increases in levels of bacterial adhesion are correlated with increased bacterial cell wall hydrophobicity. [99] Meanwhile, Herzberg et al. [100] demonstrated that the hydrophobicity of the cells is strongly related to the growth stage and with the amount of alginate they produce, by comparing a mucoid strain with a wild strain of *P. aeruginosa* in their exponential and stationary phases. The mucoid strain with higher alginate expression in the stationary phase was more hydrophilic and showed less deposition on the surface. This caused a flux decline of a reverse osmosis membrane in comparison to the wild strain, where the biofilm was grown for another five hours to reach a 30% decline in flux. Habimana et al. [101] showed that an EPS produced by mutant strain of *L. lactis* was not only highly hydrophilic, but also unable to adhere firmly to glass. The authors suggested that the EPS substances hide the physicochemical properties of the bacteria.

The EPS composition of a developing surface could, therefore, promote the adhesion of other bacterial cells to the surface, especially when the composition of the EPS matrix was changed in the period of biofilm development. Berk et al. analyzed the *V. cholerae* biofilm and showed that biofilm formation is influenced by changes in matrix composition during the early stages biofilm growth, in which an EPS protection composed of different types of polysaccharides and proteins enclosing cells was found to be capable of expanding to adjust to cell growth. [102]

Bacterial Surface Charge

Bacteria possess a surface charge due to the ionization of their cell wall functional groups, including phosphate, carboxyl and other acidic groups. [103] Bacterial cells in aqueous solutions generally have a net negative charge at a neutral pH of 7. However, the amount of the charge varies from species to species and among the strains in the same species. The charge is influenced by the age of the culture, ionic strength, growth medium, pH and bacterial surface structure. [104]

The results of the research conducted by Subramani and Hoek show that increased electronegativity of bacterial cell walls leads to reduced bacterial adhesion onto the surfaces for most bacteria. [105] van Merode et al. [106] showed great variability among strains of *E. faecalis*. They reported that the heterogeneity in cells' surface charges could increase the cells' adhesion in the initial stages of biofilm formation; however, this advantage disappeared in later stages of biofilm development. Bohinc et al. studied the zeta potential of *S.aureus, E.coli* and *P.aeruginosa* at two ionic strengths (1 mmol/L and 100 mmol/L) and found great variability among species. [72] van Merode et al.

similarly concluded that the heterogeneous strains can change the cells' wall charge in response to the changes in the environment, allowing them to adhere to surfaces with different surface properties, thereby increasing their potential to adhere to surfaces. [107] This finding could explain the advantages of mixed species biofilms in which different organisms are found in the bulk liquid, having different surface properties.

The ionic strength and the pH of the solution determine the degree of electrostatic repulsion or attraction between suspended cells. [108] Several studies have reported that the adding of ions in a bulk solution, such as NaCl and $CaCl_2$, impacts the electrostatic interactions between the surface and the cell by shielding the negatively charged surface and promoting the adhesion of negatively charged bacteria. [109-113] Chen et al. [114] found that the bacteria became less negatively charged with increased ionic strength and that neutralization of the charge is more effective with calcium than with sodium. However, the adhesion is increased with the increased ionic strength to the maximum point, at which the bacterial adhesion rate is stabilized. Chen et al. [115] also reported that the minimum and maximum ionic strengths for adhesion can vary among species. A study by Subramani and Hoek [105] showed that higher ionic strength present in bulk solutions is correlated with lower repulsion forces occurred between the bacterial cells, thus causing in higher bacterial adhesion. Van Hoogmoed et al. [116] studied the adhesion of three different strains of the same bacteria in the presence of calcium. They found increased adhesion for one strain, while the opposite was found for other two strains. Bacterial surface charge is an important factor of bacterial attachment, especially with regard to physiological conditions.

Flagella

Flagella are lash-like appendages that protrude from the some bacterial cells. Their primary role is motility, but they also often function as a sensory organelles for chemicals and temperatures outside the cell. Beside Brownian motion, several bacteria can move in liquid solution by virtue or peritrichou. Motility is general accepted as a virulence factor enhancing the colonization of target organs by pathogenic bacteria. [117] Several authors have reported that flagellar motility is important during initial cell contact to surfaces for *E.coli*, *L. monocytogenes, Y. enterocolitica*. The research emphasizes that the presence of flagellar motility enhances bacterial adhesion. [118-120] Deflaun et al. [121] analyzed the importance of flagella for bacterial adhesion and found the absence of flagella resulted in decreased attachment rates for *P. fluorescens* and *P.mutants*. Similar findings were reported by Vatanyoopaisarn et al. [120], who found that the attachment of *L.monocytogenes* to stainless steel is also affected by flagella. These studies indicate that flagella can regulate adhesion and, consequently, biofilm formation.

Surface Appendages

Fimbriae or pili are structures that are attached to the cell surface, and are defined according to their adhesive, antigenic of physical properties. [122] Type 1 fimbriae are approximately 7 nm wide and 1 μm long rod-shaped structures. They are the most common adhesins found in family of Enterobacteriaceae. [123] Some studies have shown that type 1 fimbriae have an important role in primary cell attachment for *E.coli.* [124-126] Moreover, Cookson et al. [127] showed that fimbriae are also important for Shiga toxin-producing strains.

In addition to type 1 fimbriae, other types of fimbriae have been shown to determine biofilm formation. Di Martino et al. [128] showed that *K. pneumoniae* can adhere to glass, polypropylene and polystyrene when producing type 1 and type 3 fimbriae. Another study showed that type 4 fimbriae promote the rapid formation of strongly adherent biofilms in the case of the opportunistic pathogen *A. caviae.* [129] Other surface appendices are the curli fimbriae, which are a protein source and have coiled filamntus surface structures. Ryu et al. [130] analyzed the effects of curli on attachment and biofilm formation of *E.coli* O157 H57 and determined that this is the most frequent mechanism for *E.coli* O157 H57. Some research has shown that *Streptococcus* spp. adhesion is strongly influenced by pili, and that pili have a significant impact on both invasion progress and pathogenesis. [131, 132]

Surface Polysaccharides

Most Gram-negative bacteria have exposed O-antigens on the lipopolysaccharide outer layer. This layer not only protects the bacteria from disinfectants, antibiotics and other antibacterial molecules [4], but also plays a vital role in biofilm formation. Barak et al. [133] showed that O-antigen mutants of *S. enterica* serovar *typhimurium* have reduced capacities for adhesion. Another study on *E. coli* showed that a decreased lipopolysaccharide layer did not affect adhesion directly, but had an effect on the synthesis of type 1 fimbriae and flagella, consequently resulting in reduced adherence. [134] Furthermore, the changes in the peptidoglycan structure directly at the surface of Gram-positive bacteria can also have an effect on attachment, as shown by analysis of *L. monocytogenes*. The impairment in the cellular localization of several peptidoglycan-degrading enzymes resulted in an enhanced attachment to stainless steel. [135]

In the process of initial attachment when organic matter is present, such as food scraps, blood, urine or saliva, the substratum surface becomes covered by a layer of adsorbed organic material, known as the conditioning film. [136] This film enhances the outcome of the biofilm formation process, as bacterial cells do not immediately contact surfaces. Any surface that is exposed to an aqueous medium will be coated with molecules that are adsorbed from the medium, therefore forming the conditioning film. For example, surfaces can be coated with a layer of albumin, lysozyme, glycoproteins, lipids, etc. [137, 138]

Material Surface Characteristics

In the DLVO theory, material surface characteristics are as important as cell surface characteristics. The factors determining bacterial adherence to material surfaces include chemical composition, surface charge, hydrophobicity, and surface roughness or physical configuration. [72] Moreover, the surface energy and empty binding sites can represent factors for bacterial adhesion and, consequently, biofilm formation. [139, 140]

Surface Chemical Composition

Surface chemistry composition has an important impact on bacterial attachment. [141] In a study by Cunliffe et al. [142], glass surfaces were modified with functional groups, such as amine and amides of different chain lengths showing a lack of attachment for bacteria. The study shows that *L. monocytogenes* adhesion on amine was significantly high but it was decreased by decreasing the chain length of the amide functional group. The authors demonstrated great variability among bacterial species; *L. monocytogenes* and *E. coli* showed low adhesion potential on acetamide and the opposite was found for *S. aureus* and *S. typhimurium*. Another study highlighted the importance of surface chemical composition; Faille et al. concluded that polyamides have a greater adherence potential for spores than that other polymers, such as Teflon and polyvinyl chloride. [143]

Studies in the field of water purification filters shows that bacterial cells adhere better to the material for nanofiltration and reverse osmosis membranes in comparison to membranes made of cellulose acetate. [99, 144] Moreover, bacteria tend to adhere to the cellulose acetate ultrafiltration membranes more so than on polysulphone and polyethersulphone materials. [145] In contrast, Lee et al. [146] studied reverse osmosis membranes comprising a polyamide active layer and found higher adhesion in comparison to a cellulose acetate membrane. The disadvantage of cellulose acetate membranes is that the surface can be changed due to hydrolysis from microbial production. This is not known for polyamide and polysulphone membranes [147], suggesting that studies on bacteria and cellulose acetate interactions might differ from those on bacteria and polyamide. [148, 149] This indicates that chemical composition is not the only factor governing bacterial adhesion on the membranes.

Surface Roughness

Surface roughness can be defined as a surface of peaks and valleys with various heights and spacing in a plane. The profile of a surface with submicron features is measured with an atomic force microscope (AFM). [150] The most frequently studied roughness parameter is the roughness average (R_a) that defines the typical height variation of the surface. [151]

One of the reasons rough surfaces are more easily colonized is due to diminished shear forces and the fact that the relative surface area is increased. [72, 152] However, some studies have also proven that some bacterial species colonize smooth surfaces equally well as they do rough ones. [12] It is generally accepted that the smoother a surface is, the lower the adhesion rate of bacteria is present. [72] Whitehead and Verran reported that stainless steel with low gradation has less adhesion than that with high gradation. [153] Wu et al. analyzed the coverage of titanium surface (R_a = 0.83 mm), grit-blasted surface (R_a = 11 mm), polished surfaces (R_a = 0.006 mm) and plasma-sprayed titanium (R_a = 33 mm) by *S. epidermidis* and found no significant correlation between roughness and adhesion. More insightful research using SEM has shown that this strain tends to adhere to grooves and depressions whose dimensions are similar to those of bacteria. [154] In contrast, Taylor et al.

[155] studied the adhesion of *P. aeruginosa* and *S. epidermidis* to polymethylmethacrylate surfaces within a large range of roughness (R_a between 0.04 and 7.89 mm). They found that smallest increase in roughness almost doubled the bacterial adhesion on the surface.

Several authors reported that that bacterial adhesion is promoted when the properties of the surface have dimensions or spacing similar to the size of bacteria. [153, 154, 156, 157] Some nanoscale studies showed an enhanced bacterial adhesion on nano-rough surfaces whereas others showed the opposite. [158-160] Crawford et al. [151] proposed that two variables should be taken into the account when roughness is assessed. One is a spatial parameter that describes the distance between two points of the same height, and the second is hybrid parameters that combine amplitude and spacing parameters.

Another study by Truong et al. [158] showed that the adhesion on titanium surfaces by bacterial cells *P. aeruginosa* and *S. aureus* is promoted by the presence of nanoscale topographical properties. Similar results were found by Whitehead et al. [161] in their analysis of bacterial adhesion on nanostructured titanium surfaces; they proved that the colonization is increased by increasing surface roughness. Furthermore, Bakker et al. [162] reported similar results for polymer surfaces with nanometer scale roughness. Studies on the surfaces of contact lenses have demonstrated enhanced adhesion to surfaces with greater surface roughness. [163, 164] Results of the study by Singh et al. [165] show that an increase in the surface pore aspect ratio and volume is related to the increase of surface roughness for bacteria *E. coli* and *S. aureus*. Giraldez et al. [163] found that larger surface roughness determines a greater specific surface area, thus providing more available active surface for adhesion processes. Fan et al. [166] reported evidence that a rougher surface enhanced bacterial adhesion. One explanation could be that the bacteria are protected against shear forces by surface irregularities. Nonetheless, the roughness of the surface increases the contact area of the cell with the surface. Most of the studies reports similar effects of promoted bacterial adhesion and colonization on rougher materials. [164, 167-169] However, there are also other studies that have shown that the surface roughness has no significant impact on bacterial adhesion. [96] Furthermore, Li et al. [170] reported that there was no significant effect of surface roughness on the adhesion of *P. aeruginosa* and *E. coli*.

Surface Morphology or Configuration

Surface morphology is another parameter that defines the surface properties of the material. It is defined as a morphological description of the three-dimensional surface pattern, such as a monofilament surface, a braided surface, a porous surface, or a grid-like surface. [171-173] The surface morphology can be determined in practice by SEM. The irregularities of material surfaces enhance the cell adhesion and consequently the biofilm deposition; meanwhile, smooth surfaces with no or fewer irregularities do not allow bacterial adhesion and biofilm deposition. [174, 175]

Bacteria tend to more efficiently colonize more porous material. Merritt et al. [176] analyzed the infection rates of porous and dense dental materials and found a much higher rate for porous material. Similarly, Cheng et al. [177] reported that a porous ceramsite carrier is efficient for the treatment of coking wastewater in a biofilm reactor. Porous materials can be applied to medical implants containing antibacterial agents for the prevention of biofilm. [178, 179] This indicates that bacteria better adhere and colonize the porous surfaces. [176] Peters et al. [180] analyzed intravenous catheters and the adhesion of *Staphylococci* and found that the initial bacterial adhesion occurred at different irregularities on the inner surface of the catheters.

Surface Hydrophobicity

Hydrophobic surfaces are extremely difficult to make wet. The contact angles of a water droplet exceed 150°, which is also referred to as the "lotus effect," after the super hydrophobic leaves of the lotus plant. One way to characterize the material surface is by hydrophobicity and surface charge. Metallic materials are generally hydrophilic and negatively charged while for polymers these characteristics depend on their composition. Surface hydrophobicity or hydrophobicity are determined by surface wettability, which can be measured by the contact angle. [181, 182] Van Der Mei et al. [183] reported that hydrophobicity is an important factor of bacterial cell adhesion. In the literature, it is generally accepted that most bacteria tend to adhere more efficiently to hydrophobic materials. [98] One exception is *L. monocytogenes*, which seem to form biofilms more rapidly on hydrophilic than on hydrophobic surfaces. [184] In theory, attachment between two hydrophobic elements is limited because their repulsive forces. Adhesion to hydrophilic materials is more efficient than to hydrophobic ones because the surface tension of the bacteria is greater than that of fluid media. The opposite phenomenon might occur when the surface tension of the suspending liquid is greater than that of the bacteria. [185]

Most metal surfaces possess high surface energy, are negatively charged and hydrophilic, while polymers such as Teflon have low surface energy, have less charge and are hydrophobic. When analyzing hydrophobicity, both material and bacterial cell hydrophobicity have to be determined to predict the adhesion of cells. Several authors reported that hydrophilic surfaces do not promote bacterial adhesion and *vice versa* for hydrophobic materials. [78, 104, 186] Fletcher et al. [187] conducted research in which different materials were exposed to bacterial bulk solutions. They found that bacteria tended to attach to hydrophobic plastics with no surface charge; moderate adhesion rates were found for hydrophilic metals with a positive or neutral surface charge; very few attached cells were found on hydrophilic materials. Another study by Satou et al. [188] showed adhesion of *S. sanguis* and *S. mutans* bacterial cells to the four modified glass surfaces with different hydrophobicities. They found that *S. sanguis* strains attach more efficiency to more hydrophobic glass surfaces than others.

Al-Makhlafi et al. [189] analyzed the impacts of protein coatings, such as bovine serum albumin, bovine glycoprotein, and fatty-acid free bovine serum albumin, and found that protein-modified surfaces decreased surface hydrophobicity, causing reduced bacterial adhesion to the surfaces. In another study of hydrophilic surfaces, the adherence of *S. mutans* in the presence of saliva showed low attachment to the surface. The study of the impacts of surfactants on bacterial adhesion of *S. epidermidis* included a series of 16 pluronic surfactants. The results show significant reductions of bacterial adhesion. One explanation can be that surfactants form a coated film with adsorbed chains, causing nonspecific anti-adhesive effects that can limit bacterial attachment. [190] Similar effects of polyelectrolytes were also found by other researchers. [191-194]

Conditioning Film

Polymers from a solution will eventually coat any material surface exposed to an aqueous solution. The chemical modification of the surface will affect the adhesion rate and promote biofilm formation (Figure 5). Loeb and Neihof [195] were the first to report the formation of these conditioning films on surfaces exposed to seawater. They found that films are of organic origin and that they start to form within minutes of exposure, and continued to grow for several hours.

The chemical composition of conditioning films may be different for exposed surfaces in the human environment. An example is the protein-conditioning film called "pellicle," which develops on tooth surfaces in the oral cavity. Pellicle is made of proteins as albumin, lysozyme, glycoproteins, phosphoproteins, lipids, and gingival crevice fluid. [196] As a consequence of pellicle, bacteria from the oral cavity can colonize the surfaces more efficiently.

Researchers have found conditioning films not only in the oral cavity, but also in other human cavities and fluids. For example, Mittelman [197] noted that a number of conditioning films, such as blood, tears, urine, saliva, intravascular fluid, and respiratory secretions, enhance the colonization of bacteria to biomaterials. In another study, Ofek and Doyle [198] also found that the surface energy of the bulk fluid may affect the hydrodynamic interactions of microbial cells with surfaces by altering the surface characteristics.

Reversible Attachment

After short period, a surface exposed to the bulk solution will be deposited with bacterial cells. Conditioning film on the surface will promote the adhesion of cells. After the cells are deposited on the surface, some of them will be released into the bulk solution by the desorption process. The cells that remain on the surface are irreversibly attached (Figure 5).

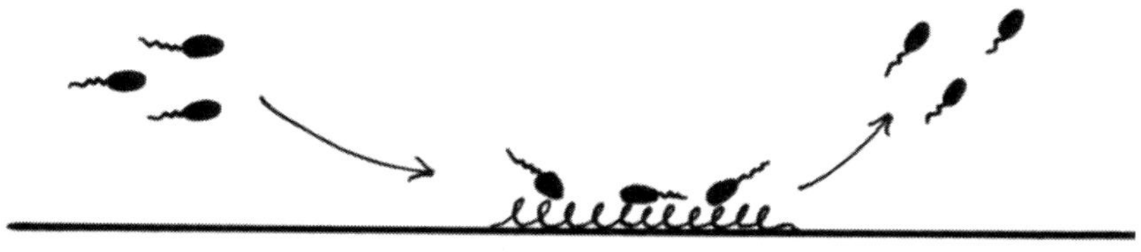

Figure 5. Bacterial cell deposition, adsorption and desorption on surface.

Irreversible Attachment

When adhesion of bacteria is changed from reversible to irreversible, attachment causes strong and permanent bonding to the surface with the help of EPS (Figure 6). [199] Gerke et al. [200] reported that *S.epidermis* started to produce EPS immediately after adhesion, forcing the cells together and promoting biofilm maturation. When the irreversible attachment is achieved, significantly stronger shear forces or chemical breaking of the attachment forces by enzymes, detergents, surfactants, sanitizers [201] and heat are needed for removal of the biofilm. [152, 200,202-206]

CONCLUSION

Bacterial adhesion is an extremely complex process, and most of the processes involved take place simultaneously, independent of each other. [207] In the process of adhesion, as the initial phase of biofilm formation, properties of bacterial cells and material surfaces play a vital role. [208-210] Typically, solid materials in various environments do not have exposed bare surfaces, and various substances are adsorbed to the surfaces before the microorganisms, thereby forming conditioning films. [80, 137, 211-214]

Some research shows that proteins on outer microbial cell membranes are heterogeneous; many of them protrude from the cells, forming cell appendages. Moreover, different structures on the same cell affect the net cell character, but they are often analyzed in terms of their synergistic contribution to the net cellular properties. [215]

Chapter 3

BIOFILM LIFECYCLE

"The main facts in life of every being are birth, food, love and death"
(E.M. Forster)

After irreversible attachment, bacteria start to multiply and to excrete the EPS that protects them from non-ideal environmental conditions (Figure 6). With the development of biofilms from a young to a mature state, bacteria start to communicate with each other, assuring that the biofilm acts as an autonomous architectural structure. [16, 216, 217] In a mature biofilm, some internal and external influences can release the cells that can return to the bulk flow. Biofilm dissemination is believed to be crucial for the biofilm lifecycle.

LIFE ON THE SURFACE

Bacterial cells can colonize every kind of material, from stainless steel to glass and wood. Azaredo and Oliveira found that the EPS produced by *S. paucimobilis* possess surface properties that assist bacteria in their attachment to hydrophilic surfaces. [218] Interestingly, the presence of adsorbed proteins on a surface prior to inoculation generally reduced the adhesion of *L. monocytogenes* regardless of the surface composition or free energy. [189] EPS have been shown to enhance nutrient capture and resistance to environmental stress and antimicrobial agents. [219, 220] When mature, biofilms exist as structured matrices with a network of vertical and horizontal channels to allow liquid to flow and thus to guarantee a supply of nutrients and the disposal of waste products that are generated as part of the natural respiration activities of cells. [71]

Exopolysaccharide Matrix

As mentioned in the previous chapter, EPS are one of the key factors responsible for binding cells and other particulate materials together by impeding the mass transport of antibiotics through the biofilm, probably by binding directly to these agents. [10, 51, 208] Several authors reported that the main components of EPS are polysaccharides, proteins, nucleic acids, lipids, phospholipids, and other organic substances. [221, 222] Tsuneda et al. [223] reported that proteins and polysaccharides account for almost 90% of the biofilm EPS structure, indicating that they are the major components. The EPS matrix has a role as a physical barrier in which diffusive transport prevails over convective transport, which is significant for bulk solutions. The most important function attributed to EPS is their vital protective effect on biofilm against adverse environmental conditions. One of these conditions is the presence of antibacterial substances that cannot penetrate the EPS or can have only limited effects. Therefore, fully matured biofilms can cope with high concentrations of antibacterial substances. [32, 224] Moreover, the EPS matrix is responsible for communication, and community behavior may accumulate at concentrations high enough to be effective. [225] The role of other EPS components (e.g. polysaccharides and proteins) are not totally understood. [226] One exception is bacterial alginates, which have been well studied with regard to their relevance as a virulence factor in infection as well as in the context of their potential commercial applications. [226] Lipids and nucleic acids might have significant influence on the stability of biofilms and mechanical properties of EPS. [227] Whitchurch et al. [228] proved that extracellular DNA is required for the initial establishment of *P. aeruginosa* biofilms, and possibly for other biofilms that specifically release DNA.

The EPS matrix structure and composition is strongly influenced by bacterial strains, physiological conditions, nutrients and gasses. [50] If there are a sufficient concentrations of EPS, the biofilm begins to act in a gel-like matter. [229]

The main substances of EPS are polysaccharides. In the case of Gram-negative bacteria, the polysaccharides are neutral or polyanionic. The presence of uronic acids (e.g. as D-glucuronic, D-galacturonic, and mannuronic acids) causes the anionic property. [217] These findings are important due to the association of divalent cations, such as Ca^{2+} and Mg^{2+}, which can provide a greater binding force in a fully matured biofilm. [230] In contrast to this, in the case of Gram-positive bacteria, the chemical properties of EPS may be different due to their cationic origin. Hussain et al. [231] reported that the EPS of coagulase-negative staphylococci have the character of a teichoic acid and proteins. EPS is, due to its highly hydrated nature, able to (due to hydrogen bonding) incorporate large amounts of water into the structure. Some researchers have reported that the EPS can be hydrophobic, but most bacterial EPS are both hydrophilic and hydrophobic [222, 232]; moreover, the EPS can vary in solubility. [85, 100]

The composition and structure of the polysaccharides define their primary conformation, since many bacterial EPS possess backbone structures that contain hexose residues that are more solid, less deformable, and poorly soluble or insoluble. Furthermore, the EPS of biofilms are not generally uniform in composition but can vary spatially and temporally. [217] Leriche et al. [233] showed that different bacterial strains produce differing amounts of EPS and that the amount of EPS is increased with the maturation of the biofilm.

Sutherland [217] analyzed the impacts of different nutrients on EPS formation and showed that EPS production is significantly affected by nutrient status, such as the availability of carbon, and limitations in nitrogen, potassium, or phosphate. Moreover, the rate of bacterial growth influences the EPS since slow growth of bacterial cells promotes EPS production. [234]

Early Development of Biofilm Architecture

The early development of biofilm starts with micro-colony formation due to bacterial adhesion to surfaces and the production of EPS [235], which strengthens the adhesion between the bacteria and the surface, and strengthens the colony against environmental conditions (Figure 6). [152]

Figure 6. Micro colony formation.

Some studies have shown that micro-colony formation could incorporate the planktonic cells from the surrounding bulk solution as a result of so-called “quorum sensing” (QS) or cell-to-cell communication. [236] O’Toole and Kolter [237] have shown that one of the conditions for micro-colony formation in the case of *P. aeruginosa* is the presence of type IV pili. In addition to the pili expression, motility was also considered to facilitate the bacterial cells to overcome the repulsive forces of the surface water interface, allowing them to reach the surface and form micro-colonies. [238] The micro-colonies represent a vital step in biofilm formation, as a basis for biofilm development and function. [239]

Maturation

The second step of the biofilm lifecycle is growth to the point that it develops into an organized structure, which can have a flat or mushroom shape, depending on the bacterial strain and environmental conditions. [235, 240] Since bacteria grow at different rates, the time required to reach maturity depends mainly on the strain. Some authors reported that a biofilm achieves structural maturity in ten days or more. [199] Micro-colonies grow from a sessile form into diverse, complex enclosed micro-colonies with open water channels. [241] The study by Whiteley et al. analyzed the DNA of mature biofilms of *P. aeruginosa* and compared it with that of bulk cells. The authors reported that over seventy genes were altered, e.g. genes encoding proteins, genes responsible for translation, metabolism, membrane transport and secretion. [21]

The maturation process starts when the extracellular matrix increases around the micro-colonies due to continued secretion of EPS. Maturation is strongly influenced by QS in which bacterial cells can communicate and regulate population growth and gene expression (Figure 7). Gram-negative bacteria can communicate with each other based on molecules of N-acyl-L-homoserine lactone, while Gram-positive bacteria do so based on auto-inducing peptides, and both groups of bacteria can communicate based on autoinducer-2. [242, 243]

When the biofilm matures, cavities are developed for the transport the water, nutrients, planktonic cells and waste. [20] Figure 7 shows the diffusion of different molecules from the bulk solution to biofilm and *vice versa*. Nevertheless, bacterial cells, located deeply in the biofilm, experience a nutrient limitation, causing the structural segregation of bacterial cells. Bacterial cells close to the surface reproduce in low rates. Since the majority of antimicrobial substances affects the dividing of cells, this can explain one of the resistance mechanisms. [3] Some research indicates that nutrient availability is a vital factor impacting the biofilm metabolic activity and that a lack of nutrient causes biofilm maturation by cell speciation. [244-246] The nutrients are available in the bulk fluid or are divided between the aqueous and solid phases, such as minerals and organic matter. In addition to the total amount of nutrients, the ratio of different nutrient components in a bulk solution also defines the limiting factors. [247] The transport of nutrients into the biofilm, as well as of metabolites out of a biofilm is defined by the flow rate of the bulk liquid and by molecular diffusion. [248]

Due to the increase in the bacterial cell mass in the biofilm formation, the entire biofilm becomes metabolically and physiologically stratified. [249] Because biofilms grow under diffusion-limited conditions and of different diffusions of nutrients, the growth rate becomes highly varied between cells that are exposed to the large amounts of nutrients at the top of the biofilm and those close to the surface. [250] Therefore, concentration gradients develop within different sections of the biofilm. The concentration gradient is determined by the biomass growth and activity and the net nutrient flux from the bulk liquid. This indicates that the relation between the physical, chemical and biological interactions is one major cause of biofilm heterogeneity, involving the chemical gradients' bacterial cell properties in different physiological states. [21]

Figure 7. Maturation of biofilm.

Some authors reported that when bacteria come into contact with surfaces, this triggers gene expression changes and factors such as those responsible for the formation of the EPS. [251-253] Several authors have found that cellulose is an important element of *E. coli* pellicle biofilms. Moreover, cellulose can be expressed with curli in some strains of *E. co*li. [254-256] Other studies have proven that polyglucosamine and colanic acid impact biofilm architecture [257, 258], with polyglucosamine being the most important of the clinical isolates of the *E.coli*. [259]

In contrast to *E.coli*, more detailed research on EPS matrices have been ongoing in the case of *P. aeruginosa* and have been shown to vary depending on environmental conditions. [260-262] Several studies have shown that extracellular DNA is important for binding the cells and firming the *P. aeruginosa* biofilms. [228, 263] The early stages of *P. aeruginosa* biofilms are more sensitive to treatment with the enzyme DNA-ase in comparison to fully mature biofilms. This can be explained by the extracellular DNA having the role of stabilizing the new biofilm when EPS components are not abundant. [228] In the process of biofilm maturation, extracellular DNA amounts are increased due to the lysis of the bacterial subpopulation. [264] The extracellular DNA in biofilms has a well-organized structure and supports the base of the mushroom-shaped biofilms. [264] Similar results were also found by Barken et al. [265], indicating that type IV pili have a great affinity to bind with extracellular DNA and consequently enhance the migrating of bacteria cells toward locations of extracellular DNA.

The significance of extracellular DNA to biofilm formation has also been reported for *E. faecalis*. [266] Researchers have also reported that several enzymes have impact on autolysis and DNA release. [266, 267] Mohamed et al. [268] reported that a lack of the autolysin reduces biofilm formation. Another study also reported that autolysis plays an important role in the regulation of DNA secretion at different stages during biofilm formation. [269]

Dispersion

Dispersion of the biofilm is the final step in the biofilm formation cycle, and it allows the bacterial cells to return into their planktonic bulk solution (Figure 8). [270] Several different environmental factors can trigger biofilm dispersion, one of which is increased fluid shear caused by external perturbation [271], but also internal biofilm processes, such as enzymatic degradation, the release of EPS or surface-binding protein. [272, 273]

It is important to realize that biofilm growth on surfaces is always a dynamic process. The last stage of the biofilm lifecycle is the dissemination of bacterial cells through liquid flow. [274] Some researchers have pointed out that that the spatial limitations of the biofilm's towers of cells and extracellular matrix in combination with a liquid flow causes a structural failure, and that the bacterial cells leave via erosion. [275, 276] Other researchers have reported that cell dispersal occurs due to quorum sensing. Wilson et al. [277] reported that is more likely that both principles are present, and that the exact type of biofilm dissemination will be dependent on the type of bacteria and the environmental conditions. They found that both phenomena appear in case of *P. aeruginosa*.

There are five main mechanisms by which bacteria can disseminate from biofilms: abrasion, grazing, erosion, and sloughing. [278] These types of biofilm detachment have been described from the aspect of the physical environment acting upon a biofilm. [279, 280]

Figure 8. Dispersion of biofilm.

Abrasion

Abrasion is defined as the cells detaching from a biofilm because of bulk liquid collision particles. [281, 282] Eventually, all biofilms under flowing conditions will experience abrasion phenomena. The process depends on the fluid flow and on the size and energy of suspended particles. Examples of systems in which abrasion is a major factor are sand filters, milk pipelines and blood. The numerical model of So et al. [283] demonstrated that low detachment forces caused abrasion in the outer pores, which limited the growth of biofilms. Meanwhile, high forces on the surface enabled deeper oxygen penetration with increased biomass activity.

Figure 9. Biofilm abrasion process.

Grazing

Grazing is a process of biofilm detachment by the feeding activity of eukaryotic organisms (Figure 10). Different microorganisms can cause the dispersion by feeding themselves with biofilm biomass. This mechanism can substantially decrease the amount of biomass, depending on the number of grazing organisms and their efficiency. Grazing by protozoa is a common phenomenon in bacterial dispersion. Sun et al. [284] have shown that wild types of *V. colerae* are resistant to grazing; one of the reasons may be quorum sensing. Similar results were also reported by Crutsinger, when testing *S. marcescens.* [285]

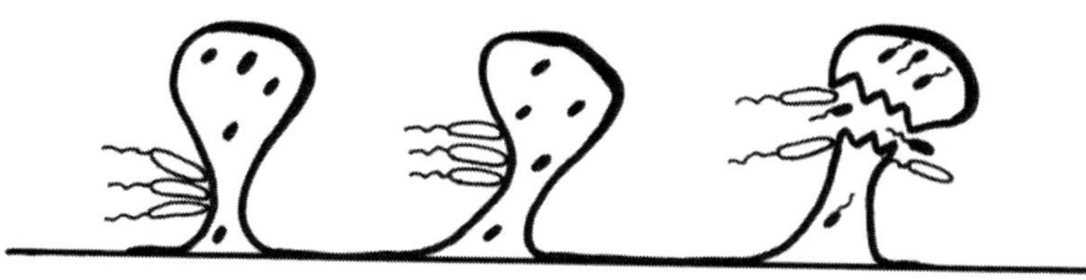

Figure 10. Biofilm grazing process.

Erosion

Erosion is defined as continuous depletion of small amounts of biofilm due to fluid shear in a flowing system (Figure 11). The cells that are closest to the bulk water forces are the most exposed to fluid shear. [286] The most sensitive are young cells and cells that are not fully incorporated. Erosion can also appear as a result of the shearing effects of the interface between gas bubbles and biofilms immersed in bulk liquid. Bester et al. studied the erosion of *Pseudomonas* spp. under fluid flow and showed that increasing forces on biofilm structures will detach single cells from the biofilm. [287] Ymele-Leki and Ross [288] demonstrated erosion on *S. aureus* biofilms under physiologically relevant fluid shear forces and emphasized that removed cells can grow on a new location.

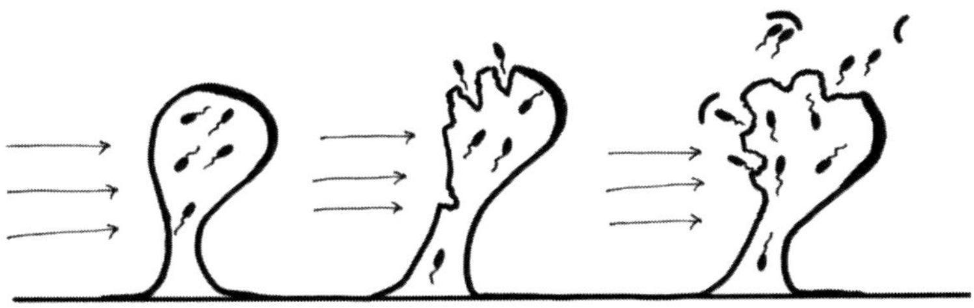

Figure 11. Biofilm erosion process.

Sloughing

Sloughing means the removal of the biofilm by fluid frictional forces. In this case, large amounts of biofilm pieces, containing large clusters of bacteria cells together with EPS, are detached. [289] In most cases, sloughing does not involve the removal of cells attached directly to the surfaces, although this can occur when attachment forces between the bacteria are greater than the attachment to the surface material. [290] Moreover, sloughing causes total detachment of the biofilm. In some cases, it is difficult to define which process is present, since differences between sloughing and erosion may not be easy to determine because there is a continuum of detached particle sizes, from very small to very large, in many biofilm systems. [71] Kumpel and Nelson [291] found sloughing for *E.coli* biofilms in water pipelines when a biofilm was exposed to pressures between 10 and 17 psi. They also demonstrated the total elimination of a biofilm after chlorination and exposure to pressure above 17 psi. In another, study Wang et al. [292] showed that the starvation of *E. coli* biofilm on sand material resulted in the sloughing and dissemination of bacterial cells.

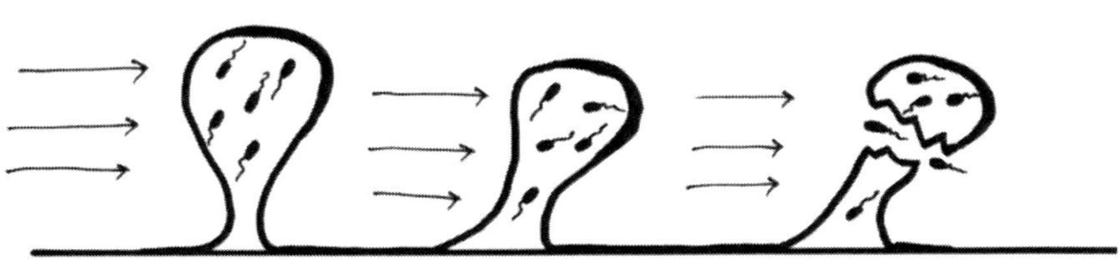

Figure 12. Biofilm sloughing process.

Rolling

Rolling of biofilm appears as a consequence of the impact of fluid flow when biofilm pieces are moved along the flow (Figure 13). [293-295] Rupp et al. [293] published time-lapse images of *S. aureus* biofilm pieces slowly moving along with the liquid flow rolling across the surface. The authors reported that the rolling of biofilm particles resulted in multiple attachments in the direction of flow. Weak rolling of *E. coli* biofilm on the surface, thus allowing more efficient spreading during growth, was demonstrated by Anderson et al. [294]

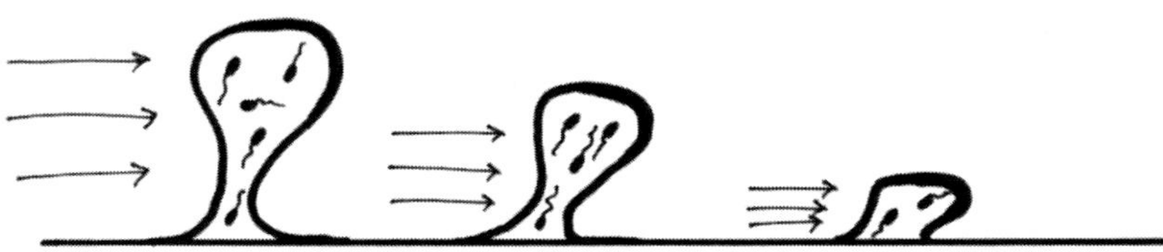

Figure 13. Biofilm rolling process.

Results from a number of studies have shown that a large number of parameters may influence detachment, including temperature, the presence of enzymes, gas bubbles, shear force and disinfectants. [296-301] Detachment is an active process with the consequence of biofilm dissemination and bacteria being deposited in new niches. [270, 302] Not only physical forces, but also starvation is considered to be a cause of detachment, which allows bacteria to search for new environments. [16, 303] Some researchers have reported that enzymatic activity can cause EPS degradation, resulting in biofilm dissemination. Kaplan et al. [273] found that the enzyme dispersin B degrades the *S. epidermidis* biofilm.

Biofilm Architecture

The biofilm structure, properties and function can vary greatly in time and space among species and even strains in the same bacterial species, due to external and internal processes. [304] However, common to all biofilms is the fact that they consist of a thin base film with a monolayer of bacterial cells or several layers covered with EPS. All biofilms have water channels allowing the diffusion of nutrients and the removal of waste. [199] The organisms incorporated in the biofilm structure have a significant impact on biofilm architecture. As reported by James et al. [305], biofilm thickness can be affected by the properties of bacterial cells. They found that biofilms of *K.*

pneumoniae and *P. aeruginosa* were thinner in comparison to a biofilm containing both species. One of the explanations can be that the species interact with each other, enhancing the structure of biofilm. [306] Tolker-Nielsen et al. [304] analyzed the impact of cell motility in biofilm architecture by studying the interactions of *P. aeruginosa and P. putida*. At the beginning of the experiment, each organism started to form micro-colonies. Over time, the colonies started to mix, showing the migration of cells from one micro-colony to the other. Consequently, the physical properties changed, resulting in reduced structure. The authors reported that the micro-colonies became motile. Ultimately, the motile cells dispersed from the biofilm, resulting in the dissolution of the micro-colony.

Interaction of Particles

In the structure of the biofilm, nonbacterial components can also be incorporated since bulk fluid can consist of various particles, such as food, physiological substances or environmental material. Durack [307] reported that fibrin and erythrocytes can interact with other components in forming biofilms. In the human body, this phenomenon can occur in vessels or on heart valves, where combinations of bacterial cells, fibrin and EPS were found. The problem of fibrin is that with a combination of EPS it represents even stronger protection against any hostile environment, leading to chronic inflammation.

Moreover, some researchers have found that bacteria on catheter surfaces can hydrolyze the urea and change the urine pH causing the precipitation of minerals [308-310]. Minerals and other particles that become incorporated in the biofilm can cause blockages of catheters, water systems or food pipelines. [234]

Hamadi et al. [311] found that *S. aureus* biofilm is strongly influenced by the type of milk in pipelines in the dairy industry. Lower adhesion was obtained for the case of stainless steel surfaces treated with semi-skimmed milk and the higher for skimmed milk. Similar to that, Mureddu et al. [312] reported *L. monocytogenes* biofilm in Sardinian fermented sausage processing plants.

Cell-to-Cell Communication

One of the most impressive properties of biofilms is the ability of cells to communicate with each other to develop cooperation as a strong survival strategy. [209, 313-315]

Cell-to-cell signaling has been demonstrated to play a vital role in cell attachment, growth and detachment from biofilms. [152, 274, 313] Bacteria are not solitary organisms; in fact, they are living in colonies to benefit from intercellular interactions and communications to survive in difficult environments. [210, 316, 317] In order to sense the environmental conditions and the response to them by gene expression, successful adaptation and survival strategies are vital. [313, 318]

The basis of cell-to-cell communication is auto-induction, which functions to assure the mechanisms of self-organization and regulation. [209, 319, 320] It involves an environmental sensing system that allows bacteria to monitor and respond to their own population densities. During growth, bacteria are secreting diffusible organic substances, also called auto-inducers, into the bulk solution. [321, 322] Fuqua and Greenberg [323] found that biofilms consisting of high concentrations of bacterial cells will result in high concentrations of signals and that the response to chemical signals in biofilms is a concentration-dependent process, in which a certain concentration of the signal molecule must be reached before a response is initiated. [324, 325]

Several authors have reported that different molecules are involved in cell communication: peptides and N-acylhomoserine lactones for communication within Gram-positive and Gram-negative bacteria [274, 326], while boronated diester molecules are found among both Gram-positive and Gram-negative bacteria. [209, 327, 328] Many vital processes in the lifecycle of biofilms are quorum-sensing dependent: biofilm growth, enzyme and antibiotic biosynthesis, biosurfactant production, EPS synthesis and extracellular virulence factors in Gram-negative bacteria. [47,274]

Some authors have reported that food might induce the expression the quorum-sensing molecules of many food-borne pathogens, spoilage, and non-pathogenic microorganisms. [319, 320, 329, 330]

When a certain concentration of auto-inducing molecules is reached in food, molecules might modulate the targeting gene or genes for growth and survival. [320, 331, 332] Rasch et al. studied bean sprouts and found that quorum sensing controls protease and pectinase production, as well as the siderophore production of Enterobacteriaceae and *Pseudomonas* spp. [333] Silagyi et al. found quorum-sensing molecules of *V. harveyi* in food. [334] Similar results were reported for *S. plymuthica.* [335] Another study, by Lu et al. [336], showed the expression of quorum-sensing molecules on tomatoes under refrigerated conditions. Similar environmental impacts can be observed in other studies in which quorum-sensing molecules were detected in contaminated, cold smoked salmon [337] and raw vegetables. [338]

Gene Transfer

In biofilms, the exchange of extra-chromosomal DNA is present as the mechanism of conjugation promoted in cells that are closer to each other than those in the bulk fluid. [339-344] Moreover, Ghigo [345] has reported that medically relevant strains of bacteria with conjugative plasmids develop biofilms more frequently. This author has proposed that the conjugative pilus for *E.coli* promotes the interaction between cells and surfaces in a cell-to-cell manner, indicating that the pilus has adhesion and gene transfer properties. Since plasmids can transmit information about antimicrobial resistance, this can also be a mechanism of biofilm survival strategies. Verraes et al. [346] have emphasized the pathways of antimicrobial resistance in the food chain. The genes that are responsible for resistance represent an important risk to human health, as they increase the gene pool from which pathogenic bacteria can pick up resistance traits. The authors state that food can be contaminated with resistant bacteria or their genes in three ways. The first is the colonization of resistant bacteria on foodstuff due to use of antibiotics during agricultural production. The second is resistant genes of bacteria that are intentionally added during the processing of food, e.g. starter cultures, probiotics and bio-conserving microorganisms. The third way is via cross contamination with resistant bacteria during food processing. [346]

Predation and Competition

Bacteria in biofilms can be present in the interaction with protozoa and bacteriophages, which can have a substantial impact of biofilm lifecycle. The microorganisms cooperate when cells are surrounded by identical genotypes. Since cooperation is only possible between different genotypes under non-ideal ecological conditions, different genotypes will typically compete. Competition between two genotypes of bacteria is often harmful as one of the genotype wins. However, several interactions are beneficial to one of the two genotypes; one example of this is predation. [39, 347]

Murga et al. [348] reported on the colonization and predation of protozoon *H. vermiformis* in biofilms. Similar predation mechanisms were found on contact lens biofilms of *Acanthamoeba* spp. [349] Derlon et al. [350] showed that predation influences the total amount of biofilm developed on ultrafiltration membranes. Another interaction between biofilms was noted by James et al. [305] They found competition between the biofilms of

Hyphomicrobium sp. and *P. putida* with consequences of dominance by *P. putida*. Moreover, Stewart et al. [351] proved that biofilms containing *K. pneumoniae* and *P. aeruginosa* can persist in coexistence, although *P. aeruginosa* growth rates are much lower. They also reported that *P. aeruginosa* grow as a kind of base biofilm, while *K. pneumoniae* have better access to nutrients. Another study, by Friman et al. [352], demonstrated the predation of *P.aeruginosa* with a wild type of mutant and showed that functioning quorum-sensing system confers an elevated resistance to predation.

Interactions of Pathogenic Organisms

Several bacterial pathogens have been shown to associate with and, in some cases, grow in biofilms, including *L. pneumophila*, *S. aureus*, *L. monocytogenes, Campylobacter* spp., *E. coli, S. typhimurium.* [244, 348, 353-356] Most of the abovementioned microorganisms can adhere to surfaces and on developed biofilms; most of them are not capable of extensive growth on biofilms due to specific growth conditions. [357, 358]

Several authors have shown that interaction of pathogens microorganisms with existing biofilms varies among species. One example is *L. pneumophila*, which cannot form a biofilm without free-living protozoa. [348] The survival and growth of pathogenic organisms in the biofilms might also be promoted by interactions with primary organisms in biofilms. Camper et al. [355] reported that *S. typhimurium* can grow in water filters with populations of undefined heterotrophic bacteria. Falsetta et al. [359] found that the presence of *C. albicans* EPS has an incentive effect on *S. mutans* in comparison to the absence of *C. albicans*.

CONCLUSION

The formation and growth of biofilms are dependent on many factors, including the material surface properties, bacterial strain features and environmental conditions. After the initial attachment, the bacterial cells start to secrete the EPS that protect vulnerable cells from extreme conditions. With growth, the biofilm starts to mature, and cells start to communicate with each other by quorum sensing. The biofilm starts to disseminate bacterial cells by internal and external mechanisms, ensuring the colonization of new niches.

Chapter 4

ENVIRONMENTAL CONDITIONS OF BIOFILM GROWTH

"Being defeated is often a temporary condition"
(Marilyn vos Savant)

Bacterial biofilms persist under various environmental conditions, from ideal conditions in host organisms to the extreme limits of survival. The biofilm lifecycle, from initial adhesion to the mature biofilm and its dispersal, can be exposed to the different circumstances. [37]

Environmental conditions that influence biofilms include nutrient availability, concentrations of several gasses, pH and temperature. [244, 360-364] All these conditions ultimately determine the bacterial response pattern with respect to biofilm development. Several authors have reported that in natural environments, seasonal variations and patterns can have significant influence on biofilm growth. [365-367] Water temperature or seasonally affected parameters are two such influences. Fletcher [97] reported that an increase in the concentration of several cations in bulk solutions can result in enhanced attachment of *P. fluorescens* to glass surfaces, mainly due to the reduction of repulsive forces between the cells and the glass surfaces.

Environmental conditions, such as temperature, pH or the concentration of suspended particles, have a substantial impact on the conditions of biofilms forming. [368, 369]

NUTRIENTS

Several authors have reported that the amount of nutrients is related to the increased amounts of the attached bacterial cells. [370-373] However, bacterial cells can survive and even multiply in extremely low concentrations of nutrients. [374] Bacteria in mature biofilm systems can provide nutrients by using the waste products of their neighbors and secondary colonizers, and by using their biochemical resources with different enzymes to break down food supplies. [375] Most biofilms are in a negatively charged aqueous media, causing the attraction of many nutrients and allowing the exchange of ions.

As mentioned in Chapter 3, changes in nutrient concentrations are one of the mechanisms of biofilm dispersal. James et al. [305] reported that exposing an *Acinetobacter* biofilm grown under limited conditions to high concentrations of a nutrient medium causes widely dispersed cell arrangements. Although the formation of biofilm is more frequently studied than its dispersion, some studies report that environmental factors have an impact on biofilm dispersion.

Sauer et al. [362] reported that some nutrients, such as succinate and glutamate, can promote the immediate release of cells. The authors found that dispersed cells have induced genes for motility, ribosomal proteins, and phage, while cells remain attached to the biofilm with pilus production and anaerobic nitrogen respiration. Shao et al. [376] studied *H. pylori* biofilms under starvation regimes and found that low concentrations of nutrients will suppress the proteins responsible for the flagellar movement, bacterial virulence, signal transduction and regulation.

While some studies report that rich media promote biofilm growth, some researchers have found contrary results. Kim and Frank [372] reported that *L. monocytogenes* grows better in minimal or diluted rich media. Similar results were reported by Moltz and Martin [377], showing that poor nutrients enhance biofilm formation in six out of eight strains of *L. monocytogenes*. In contrast, Stepanovic et al. [378] found that *L. monocytogenes* forms better biofilms on plastic surfaces in the presence of a rich medium as opposed to *Salmonella* spp., which produced better biofilms under low-medium conditions.

Several researchers have reported that the starvation of biofilms consequently results in inducing dispersion. [360, 376, 379] It is obvious that bacteria that experience nutrient limitation will tend to leave the biofilm system. Nevertheless, some of the bacteria grown under low nutrient levels tend to form surface-associated biofilms rather functioning as planktonic cells. [380-382] Lagha et al. [381] studied *S. enteritica* biofilm starvation over a period of three years and found that nutrient deficiency causes new adaptation strategies. They also emphasized that survival strategies may induce a genetic instability and change the virulence state. *C. albicans* biofilm grown under low nutrient concentrations resulted in low biofilm capabilities, as shown by Ning et al. [382] The authors also demonstrated that such biofilms are less complex, have a higher roughness coefficient and a higher ratio between their surface and volume.

High concentrations of nutrients can cause disruptions in biofilm growth or can result in a structure that cannot resist fluid shear. [220, 383] Dewanti and Wong [384] analyzed *E. coli* biofilm formation in different concentrations of nutrient media on stainless steel surfaces. They found that biofilms grow faster with high amounts of adhered cells in low nutrient conditions. *E. coli* biofilms were found to be thicker in a low-salt environment in comparison to the high concentrations, regardless of the carbon source. Similar findings about nutrients' impact on *E. coli* are also reported by Amato and Brynildsen. [385]

These findings indicate that bacteria under nutrient-limiting conditions tend to remain in a biofilm rather than return to the bulk liquid environment.

Shear Stress

Fluid flow can have a significant impact on biofilm formation and detachment. It is well known that properties of biofilms, such as cell density, the amount of EPS and the strength of attachment, can be affected by hydrodynamics. [247, 386] Moreover, some research has reported that biofilms grown under different flow regimes have different rheological properties. Stoodley et al. [271] found that biofilms grown under low flows form isotropic structures, while higher flows may produce filamentous cells or groupings of cells.

In hydrodynamics, the flow velocity immediately close to the surface is negligible. This zone of negligible flow is termed the "boundary layer." The thickness is related to the linear velocity; the higher the velocity is, the thinner the boundary layer is. [387] Meanwhile, areas outside the boundary layer are characterized by turbulence conditions. Flow regimes impact the adhesion of bacterial cells to the surface. Cells in hydrodynamic conditions can be defined as particles in a bulk solution, and the attaching rate to the surfaces will depend on the velocity of the liquid. [388] Bacteria that are exposed to very low fluid flow can overcome the hydrodynamic boundary layer and adhesion will depend mainly on bacterial size and motility. [207] With the increasing velocities, the boundary layer decreases, causing an explosion of the cells in response to the greater turbulence in the fluid. Increasing the fluid flow will result in higher shear forces that will increase detaching or preventing the biofilm formation. Biofilms are under the influence of the shear forces originating from the liquid flow (Figure 14). In the case of laminar flow, the shear forces are limited only to the flow direction, whereas in the case of turbulent flow, shear forces are detaching bacteria from several directions, causing a better biofilm destruction (Figure 14).

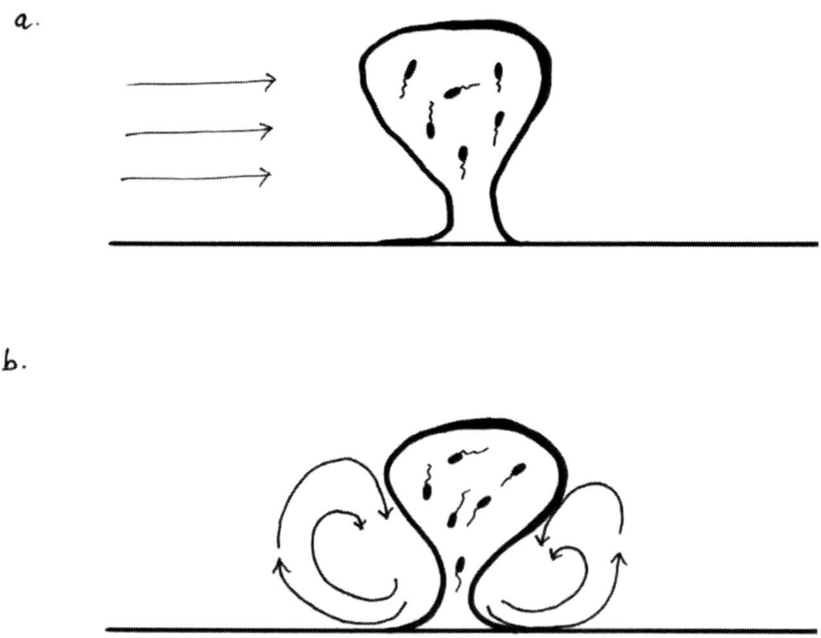

Figure 14. Impact of shear forces on biofilm under laminar and turbulent flow.

Stoodley et al. analyzed [271] *P. aeruginosa* biofilms under different flow regimes and found that biofilm grown under high fluid flow are more firmly attached than those grown under lower flow. There is also some research that hypotheses that turbulent flow may promote bacterial adhesion and biofilm formation. [12] In contrast, the detaching of entire *Staphylococcal* spp. biofilms has been observed under exposure to turbulent flow, perhaps allowing mature biofilms to colonize new surfaces. [293, 389]

Another study showed that *E.coli* adhesion to the mannose-coated surfaces is greatly dependent on fluid flow. Researchers report that at low fluid flow cells can roll downstream to the surface and that when the shear force increases, the cells become more firmly attached. [294] One explanation can be that low shear forces actually promote bacterial colonization of new niches, while under high shear forces the cells remain in tight colonies. [390] Agladze et al. [258] found that *E. coli* biofilms under a high fluid rate resulted in changes of spatial organization. Using computational fluid dynamics, Moreira et al. analyzed an *E.coli* biofilm grown under laminar and turbulent flow and found that biofilm formation was favored at the lowest flow rate because shear stress effects were more significant than mass transfer limitations. [391] Salta et al. [392] demonstrated that low shear stresses impact biofilm morphologies, forming mushroom-like structures and interstitial channels. High shear stresses compact clusters with large interspaces. Some researchers tested surfactants under shear force conditions and found the dissolution of cell attachments via surfactant production in *B. subtilis*, *P. aeruginosa* and *S. epidermidis* with significant impact on biofilm structure. [393, 394]

Laboratory experiments have shown that flow conditions have an important impact on the rheological properties of biofilms. Purevdorj et al. [395] demonstrated that the structure of *P. aeruginosa* can be changed due to the exposure to turbulent flow, causing denser and more streamlined structures. The authors also reported that under the laminar flow flat layers of biofilm were formed with rough surface topology.

Several researchers have demonstrated that low shear stress can promote bacterial adhesion to the surface during the initial steps of bacterial adhesion [396-398]; in contrast, many experimental studies have reported that shear stress acts as an inhibitor of biofilm development. Bacterial cells that are exposed to high shear forces tend to form thin monolayer biofilms [399, 400], which can inhibit maturation [379] or promote cell dispersion.

Mehdi Salek et al. [399] found that among biofilms grown under high shear stress conditions, the viability of biofilms was observed to decrease when treated with Gentamicin. One explanation for this can be that high flow rates promote molecular diffusion, and thin biofilm structures can easily incorporate the antibiotic into the matrix. Resistance to the hydrodynamic effect is time dependent and can vary from several hours to days, depending the conditions and bacterial strain. [138]

Although the stress interrupted biofilm formation and covered a large space, it assisted with the clumping and dispersion of the biofilms. Under a high flow rate, the biofilm area and thickness was reduced, but the total number of bacterial cells in the fluid was high. [401] Clumped biofilm particles migrated with the flow and provided opportunities for colonizing new niches. [389]

Fluid flow conditions can have several impacts on biofilm development. Shear stress can reduce the development of biofilm but enhances the new biofilm formation by increasing the dissemination and bacterial motility. [397, 402] The balance between these two processes can provide optimal conditions for biofilm growth. Nevertheless, it should be understood that the impact of fluid flow is specific for a single strain.

The Gaseous Environment

It is well known that biofilms can be grown under different conditions, one of which is the presence of electron acceptors, such as oxygen or nitrate. Arcangeli and Arvin [403] studied toluene degradation in biofilms grown anoxically with nitrate or aerobically with oxygen. They found that the toluene oxidation was at the highest rate in the log phase of growth with nitrate, while oxidation was highest in the lag phase with oxygen. Moreover, growth under oxic conditions was irregular and filamentous while regular and smooth with nitrate. Villaverde et al. [404] studied biofilms under oxic conditions and reported that the upper layer of the film was non-viable, but proposed that this layer protects lower layers that were respirating. Santegoeds et al. [405] took a similar research and found that that only top of the biofilm (for about 300 μm) has aerobic conditions; this allows communities close to the surface to persist in anaerobic conditions. Kroukamp and Wolfaardt [406] studied the carbon dioxide production of *Pseudomonas* spp. and found that carbon dioxide represents 25% of inflow carbon. The gaseous stratification in the biofilm structure allows a much wider range of organisms to live in the biofilm.

Several authors [407, 408] have reported that the concentration of oxygen present in the bulk solution impacts bacterial adhesion and the production of EPS. Laboratory experiments indicate that the initial attachment of *P. aeruginosa* to the surface is strongly influenced by oxygen concentration. The planktonic cells tend to adhere to surfaces in the areas with high concentrations of dissolved oxygen. [407] Skolimowski et al. [407] also proposed that biofilms grown under oxygen-limited conditions contain more extracellular polymer carbon than carbon-limited biofilms.

Researchers also show that high levels of extracellular polymer carbon promote the structural strength of biofilm and decrease erosion under shear stress. [409] Mun et al. [410] reported that dense-phase carbon dioxide was effective in inactivating biofilms that were wet but not immersed. Dense phase carbon dioxide is one potential measure to control bacteria with high antimicrobial resistance.

THE PH VALUE

The changes in pH have a significant effect on most bacteria; consequently, several detergents and disinfectants are based on low or high pH. [411-413] The antibacterial mechanism of pH is regarded as the impact on proton pumps in the membrane of bacteria. [414] The passive influx of protons in response to the proton motive force can be a problem for cells attempting to regulate their cytoplasmic pH. [415] Extremes in pH can overwhelm corrective mechanisms in bacterial cells and have an antibacterial effect on the microorganisms.

Bacteria can also adapt to the external pH changes by adjusting the synthesis and activity of proteins. [416] Serefko et al. [417] reviewed the environmental conditions of *C. albicans* on dental prostheses and reported that low pH will promote bacterial adhesion. Nevertheless, studies have shown that stepwise increases or decreases of the pH will result in the adaptation of microorganisms, while sudden changes will decrease the population. [418] However, the production of EPS is regulated at a pH of 7, indicating that bacteria cannot produce EPS matrices at low or high pH levels. [419]

TEMPERATURE

The temperature of the bulk solution is a variable that has multiple impacts on a biofilm community. Biofilm nutritional intake is temperature dependent and, consequently, so is the growth rate of biofilm. [420, 421] However, the nutrient metabolisms are directly correlated to the production of enzymes that control several processes in biofilm formation. Therefore, optimum temperatures will grow healthy biofilms; conversely, any deviation from optimum will result in structural and functional changes. [422]

Decreases in the temperature of bulk fluids can result in lower adhesion rates, as was proven by Fletcher and Pringle in the case of *Pseudomonas*. [423] One reason for this can be decreases in the bacterial surface polymer or a reduced surface area. In contrast, Herald and Zottola [424] reported that with the decreasing temperature the number of flagella in increased, which can promote initial attachment. There are several studies about the effects of temperature on biofilm growth; there is, however, a lack of research on the impact of temperature on bacterial detachment. The research of Marion-Ferey et al. [425] showed no significant impacts of high temperatures (80–90 °C) on biofilm removal. They determined that that extremes in temperature are not effective because of "baking effects" that promote bacterial adhesion. Chavant at al. [184] studied the formation of *L. monocytogenes* biofilm on metal surfaces under different temperature regimes and found that a temperature of 37 °C causes the detachment of bacteria. In contrast, Christersson et al. [426] found that temperatures in the range of 22 °C to 37 °C do not affect bacterial populations in the environment of the human mouth.

CONCLUSION

Environmental conditions in bulk solutions have an enormous impact on bacterial cell adhesion, growth, maturation and biofilm dissemination. [427-429] Therefore, controlling the conditions under which biofilms can form is one strategy for combating them. Among the conditions that affect biofilm development are temperature, pH, oxygen concentrations, hydrodynamics and nutrients. The integration of these influences determines the development of biofilms.

Chapter 5

METHODS FOR EVALUATING BIOFILMS

"Man cannot discover new oceans unless he has the courage to lose sight of the shore"
(Andre Gide)

Progress in microbiology, chemistry and physics has resulted in numerous methods to assure the reliable quantification and qualification of bacterial cells attached to surfaces as well as those in biofilms. Several methods for bacterial enumeration have been proposed by many authors [9, 72, 430, 431] including direct and indirect counting methods, such CFU (Colony Forming Units) plate count, electron microscopy and spectrophotometry for stained bacteria.

DIRECT METHODS OF BIOFILM ASSESSMENT

Light Microscopy

Light microscopy is a basic method for studying bacteria on surfaces or in biofilms. [432] The main condition for using light microscopy is that the material surfaces be translucent, which can be limitation, because stainless steel, plastic, wood and other common material are not permeable to light. Cells on the surface can be stained with dyes, such as crystal violet, carbol fuchsin, ethidium bromide, and others. Light microscopy enables a mechanism of visualizing the topology of the biofouling as well as the substrata, and aids in defining whether the bacteria are viable. Some researchers used light microscopy in a flow chamber system, which permits the study of bacterial populations on transparent material in real time. [433] Using light microscopy, Corcuera et al. [434] studied the morphological plasticity of *S. oralis* biofilms and found strain-specific growth patterns. Light microscopy can also be used to study biofilms on human surfaces. For example, Kennedy et al. [435] analyzed biofilms on burn wounds and found sepsis causes the recalcitration of biofilms. In another study, Rühs et al. [436] used light microscopy to validate the rheology and tensiometry measurement of *S. typhimurium, E. coli,* and *B. subtilis*.

Scanning Electron Microscopy (SEM)

The most commonly used method for observation of the morphology of adhered populations is Scanning Electron Microscopy (SEM) [437], the advantage of which is that it permits the characterization of the material at the same time as the quantification and orientation of bacterial cells. [175] For example, whether the bacteria tend to adhere more in scores or on peaks of material can be determined. SEM has been used for the quantification of adhered bacteria to surfaces or host tissues. [186, 438] Nevertheless, only a small part of the surface can be analyzed with this method; furthermore, sample preparation is destructive for bacteria since it involves the use of solvents and dehydration in a vacuum. [72]

SEM has been used for the study of medically relevant biofilms [439, 440] and infections. [441] Ansari et al. [442] studied the effects of jujube honey on *C. albicans* growth and found that honey affects biofilm growth and causes disruption of its structures. The authors also emphasized that 40% honey in a water solution interferes with the formation of *C. albicans* biofilms and disrupts biofilms. Similarly Lou et al. [443] used SEM for the quantification of the antimicrobial properties the burdock leaf component on *E.coli*. Brambilla et al. [444] used SEM to analyze the impact of hydrophobicity on the formation of *S. mutans* biofilms. Due to its excellent resolution properties, the electron microscope will continue to be an important tool for the biofilm scientist.

Confocal Laser Scanning Microscopy (CLSM)

While light microscopy and AFM are recommended for the enumeration of single layers of bacterial cells on surfaces, Confocal Laser Scanning Microscopy (CLSM) has some advantages in analyzing the structure of dense and hydrated biofilms. [445, 446] CLSM can analyze the biofilm morphology and physiology in four dimensions under in situ conditions. One of its advantages is that the probe is not required to be suspended nanometers from the surface, as with AFM. It is possible to scan thick biofilms by processing three-dimensional images. [447] CLSM can reduce the pre-treatments, such as disruption and fixation. Pre- treatments can reduce or eliminate information on microbial relationships, as well as biofilm organization and structures. [448] The use of CLSM with fluorescent dyes allows the determining of cells' special structures and distinguishing between live and dead organisms. [61] Bridier et al. [449] combined CLSM with the use of microtiter plates compatible with high-resolution imaging for the study of biofilm formation and structure. Danielsen et al. [450] used CLSM to study the bacterial biofilms in patients with chronic sinusitis. Meanwhile, Wu et al. [451] analyzed the antibacterial properties of silver nanoparticles against *E. fecalis* biofilm with CLSM combined with viability staining.

Atomic Force Microscopy (AFM)

AFM analysis is based on a silicon nitride tip located on the flexible cantilever. [452, 453] The tip scans the surface with a small force. Sample undulations result in the deflection of the cantilever and detection by laser on the end of the cantilever. [150] Information is transmitted to the split photo-detector and then to the piezo-scanner to form a picture of the surface. [452, 454-456] Since the 1990s, AFM has changed the characterization of microbial cells by allowing a better understanding of their walls and the mechanisms of bacterial attachment. [457] It can provide force maps to study the nano-mechanical properties of microorganisms. Moreover, it can be used to determine chemical species or specific receptors with nanometer resolution directly on the membranes of living organisms. [458] Yang et al. [459] studied *E.coli* and *S.epidermidis* with AFM after exposure to silver ions and found structural changes. *E. coli* show vesicles on the cell walls, which became larger with the increase of the incubation time. For *S. epidermidis*, deep grooves appeared on the cell walls, and the cytoplasm membrane shrank and became separated from the cell wall. AFM was also used for characterization of *E.coli* adhesion on coated glass material [460] and the characterization of antimicrobial activity. [461] Using AFM, Ribeiro et al. [462] demonstrated the effects of antimicrobial peptides against *S.aureus*. The authors reported membrane blebbing, disruption and lysis. One of the innovations in the field of bacterial imaging is atomic force spectroscopy (AFS) that combines the observation of forces on the tip as well on the sample. Van Der Mei et al. [463] reported on the interaction of the tip and *E.coli* showing that the protein on the outer membrane is related to bacterial adhesion. Several studies of medical devices were conducted with AFM. Gorman et al. [464] studied the formation of biofilm on polyvinyl chloride using AFM, and the development of resistance to ceftazime and hexetidine. In another study, Jones et al. [465] determined the surface characteristics of the oropharynx and trachea in relation to the adhesion of *Candida* spp.

Radiolabelling

The aim of this method is to label bacterial cells with radiotracer materials, such as ^{14}C leucine, ^{14}C glucose, ^{3}H thymidine or ^{3}H palmic acid. [466] It was shown that the radiolabelling has potential for the studies of bacterial adhesion to irregular material surfaces, such as inner surface of catheters, pipes or spheres surfaces. Radiolabeling can reveal structures, functions, and physiological interactions among community members in complex multispecies. Its advantages include high sensitivity, accuracy, and rapid procession. The disadvantage of radiolabelling is the exposure of laboratory staff with radioactive material. [467] Chattopadhyay et al. [468] evaluated ^{99m}Tc labelling as a potential bacteria-specific infection imaging agent for *E.coli*. Similarly, Ross and Pegram [469] radiolabelled *S. typhimurium* with a ^{35}S-label and found the radiolabelling protocol to be useful for studying DNA reactivity.

Indirect Methods of Biofilm Assessment

Staining

Spectrophotometry methods are based on the quantitative relation between the optical density of dye and its concentrations. Based on the absorbance, the colony counts can be theoretically derived from calibration curves that are specific for each bacterial strain. [72]

The use of microtiter well plates was first described by Christensen in 1985 and is commonly used for studying bacterial adhesion. [470] Although there have been several modifications of the method since then, the principle remains the same. Cells are added to the microtiter well plate and incubated. After that, the wells are washed to remove the unattached population. In the next step, any attached cells are stained with a dye, e.g. crystal violet. After the excess amount of dye is washed away, the dye that was absorbed by cells is extracted using a solvent, e.g. ethanol, acetone, and the optical density is measured at a specific wavelength. The advantage of this method is in analyzing the number of samples suitable for screening testing. Nevertheless, there are also some disadvantages, since sample preparation, solvent preparation and staining time are vital steps. However, the greatest weakness of this principle is that cannot distinguish between live and dead cells, which is especially important from the hygiene aspect. [471, 472]

Another possibility is to provide the detachment of cells from the surface with a solvent or ultrasound and then stain the cells, but the detachment process can also be problematic, since it can result in low yields, injury to the cells or cell aggregation. [473]

Plate Counting

The common laboratory method for the quantification of bacterial populations from biofilms is plate counting, in which (as with staining) attached bacterial cells have to be removed from the surface e.g. by scraping, solvents, vortexing and ultrasound. [9, 250, 474-476] Hassan et al. [477] quantified biofilm formation in the clinical isolates via a culture plate count; this was considered to be superior to both the tube method and the Congo red agar method. Similar results were also observed by Sharvari and Chitra [478] when testing *S. aureus* biofilms. Although plate counts can assure the enumeration of bacteria, they do not provide information about biofilm structure and physiology.

NOVEL APPROACHES TO STUDYING BIOFILMS

The state of bacteria at very low metabolic activity with no reproduction or the so-called "viable but nonculturable" state can represent a challenge in biofilm determination, because classic analytical methods will not work. [479-482] Examples of bacteria that can become viable but nonculturable are *E. coli, S. enteritica, S. flexneri, V. cholerae* and *V. cholera.* [483]

The study by Pasquaroli et al. [482] showed that in addition the temperature and pH level, antibiotic exposure also can induce the state of viable but nonculturable for *S. aureus* biofilms. These facts contributed to the exploration of new methods for the quantification of bacterial cells in biofilms that are viable but nonculturable.

One innovation is the Ibis technique for biofilm detection based on specific nucleotide compositions. The specially designed polymerase chain reaction amplifies a species-specific nucleotide region; the mass of the polymerase chain reaction can be quantified by mass spectrometry. [484]

Some clinical papers have reported that middle ear infection by biofilms can be detected by using light-scattering techniques, such as low-coherence interferometry and optical coherence tomography. [485] One of the greatest advantages of this non-invasive technique is to determine which areas are infected. [485]

The specific biochemical composition of bacterial cells determined by infrared spectroscopy is another option. The advantage of this method is that it can provide results in one hour and can replace some of the traditional methods in food microbiology. [486]

For detecting biofilms on surfaces in food sanitation and production, hyper-spectral imaging technology can be used. Common biofilms, such as *E. coli*, *Salmonella* spp. and *Pseudomonas* spp., can be quantified due to the production of fluorescence emissions when excited by ultraviolet light. [487]

Some of the methods that can be used for early biofilm detection are acoustic and cantilever sensors that produce currents for biofilm determination. [488]

Mathematical Modelling

Mathematical modelling aims to understand biofilm structures and physiology and can be used not only to determine results, but also to make predictions. [489] In the last ten years, progress in computer fluid dynamics and microbiology has enabled making accurate models with high prediction reliability.

Some of the models were successfully applied to explain and predict the biofilm's behavior in various conditions. The models are based on equations of mass conservation, temperature, concentration and flow. They can predict the growth, dynamics and spatial distribution of biofilms. [490, 491] Most authors report that the fluid flows of bulk solutions have a significant impact on shaping the structure of the biofilms through the nutrient convection and the detachment of biomass by shear stress. [13]

Based on the previously established models and a better understanding of biofilms' properties obtained through experimentation, several biofilm models taking into account bulk fluid have be developed, numerical simulations in 2D and 3D have been carried out. [37, 492]

CONCLUSION

Biofilms are complex communities of microorganisms; therefore, analyzing a biofilm's structure, organization and functions requires multiple approaches for adequate characterization. The methodology of biofilm quantification and qualification has shifted from classic methods to the more interdisciplinary approaches. Each method has its advantages and disadvantages. It is evident that the most significant methods have to take into the account the chemical, biological, ecological and physical issues related to biofilm analysis.

Chapter 6

Biofilms in the Food Industry

"Happiness and bacteria have one thing in common; they multiply by dividing"
(Rutvik Oza)

Foodborne diseases are illnesses resulting from the consumption of contaminated food. [346, 493, 494] Such contamination may occur in any of the stages in the "farm to fork" chain. The World Health Organization stated that foodborne diseases are considered to be emergent public health issues. Moreover, the Centre for Disease Control and Prevention reported that in the last decade foodborne diseases have been related to the more than 1000 reported disease outbreaks [495] with the 48 million illnesses, 128,000 hospitalizations, and 3000 deaths annually in the United States. [496] Several authors have found that many foodborne outbreaks are related to the formation of biofilms. [497-502]

In recent foodborne outbreaks, searches for the source of food contamination have focused on the detection of biofilms. [503-505] Biofilms are a significant issue in the food-producing industry since they are more resistant to sanitation and disinfection than planktonic cells. [506-508] Biofilm control has distinct issues in a wide range of food industries, including brewing [509-513], seafood processing [514-520], dairy processing [235, 521, 522], and meat processing. [505, 523-525] Biofilms in the food industry are found on a large range of materials, such as glass, stainless steel and polystyrene. The most common bacteria isolated from food contact materials are *Salmonella* spp. *E. coli*, *P. aeruginosa* and *S. aureus* (Table 1).

Table 1. Common bacteria forming biofilms on food contact materials

Biofilm type	Contact materials	References
Salmonella spp.	glass, polyvinyl chloride, stainless steel	De Oliveira et al. [526] Ban et al. [527] Grough and Dood [528] Helke et al. [529] Lagha et al. [381] Proutry and Gunn, 2003 [530] Nguyen et al. [531]
E. coli	stainless steel, glass	Rivas et al. [532] Posada-Izquiero et al. [533] Visvalingam and Holley [534] Mitik-Dineva et al. [160]
P. aeruginosa	stainless steel, glass	Dargahi et al. [298] Simões et al. [206] Moritz et al. [535] Cloete and Jacobs, 2001 [536] Bourion and Cerf, 1996 [537]
L. monocytogenes	polyethylene, wood	Goh et al. [538] Adetunji and Isola [539] Perni et al. [540] Ban et al. [527]
S. aureus	polypropylene, stainless steel, Teflon, ceramic	Kouider et al. [169] Oulahal et al. [204] Li et al. [541] Hamadi et al. [542] De Souza et al. [301]
B. cereus	stainless steel, ceramic, polystyrene	Klavenes et al. [543] Hornstra et al. [544] Kumari and Sarkar [545] Li et al. [541]
Campylobacter spp.	stainless steel, polystyrene, glass, polyurethane	Nguyen et al. [546] Sanders et al. [547] Gunther IV and Chen [548] Nguyen et al. [549]

Raw Vegetable Produce Industry

Bacterial cells can contaminate various type of food when they are protected by plant stomata, cracks and cut surfaces, so that they can produce biofilms on surfaces. [550, 551] In the process of making fresh produce, steps including trimming, cutting, washing, rinsing, dewatering and packaging are known to be primary sources of cross-contamination and, therefore, the most important cause of food-borne illnesses. [125, 493, 552] An essential study of López-Gálvez et al. [553] showed that washing with chlorine or sodium hypochlorite has no effect on bacteria inactivation after contamination of fresh cut lettuce with *E.coli*. A more detailed analysis with SEM showed that cells adhered to the lettuce stomata, where they were protected against disinfectants.

Several authors have pointed out that the aim of the disinfectants is to prevent cross contamination from the source to the product. [553-555] Bonsaglia et al. [556] detected *L. monocytogenes* on polystyrene, glass and stainless steel materials. One of the critical control points in the process of producing fresh products is the packing intended to prevent cross contamination. Produce is generally consumed raw; therefore, packing as the last step of processing is considered to be a critical control point. This step should be monitored and controlled attentively and regularly to avoid re-contamination. Carron [557] reported on an outbreak with *L. monocytogenes* that was connected to the unhygienic packing of melons.

Some of the studies reported the colonization of bacteria on conveyor belts, drying areas and floor drains. [9, 558] Bacteria can colonize a vast spectrum of surfaces. When the cells are irreversibly attached to the surface, they form a biofilm that can contaminate the entire process line or even the entire production facility. [559] Microorganisms tend to attach themselves to intact surfaces of produce. [560] Bacterial cells can penetrate the food surface and form biofilms inside of the food. [561] In a study by Han et al. [562], it was found that there is no significant colonization of *E.coli* on intact green peppers while significant growth was found on damaged ones.

In the food industry, sanitizers such as ethanol, ozone, chlorine, and organic acid have been found to be ineffective against biofilms. [563] In addition, *E. coli* was reported to be resistant against a 20,000 ppm chlorine, which is much higher than for planktonic cells. [564]

Another study also showed that a 1% hydrogen peroxide solution is ineffective in *E. coli* removal on melons; it was observed to be reduced by less than 1 log. [565]

DAIRY INDUSTRY

Bacteria in the dairy industry often contaminate pipes, vessels and other equipment. In addition, the milk itself is highly vulnerable to bacterial contamination and can be quickly spoiled. In the dairy industry, the most significant sources of milk contamination are considered to be unhygienic equipment. [566] The common bacteria that are found in dairy products belong to the genus *Enterobacter* [567, 568], *Listeria* [330,569-573], *Streptococcus, Bacillus* [569, 574-576] and *Pseudomonas.* [577-580]

Latorre et al. [570] analyzed the sources of *L. monocytogenes* in dairy farms and found biofilms on scratched surfaces. Helke et al. [529] conducted a study of *L. monocytogenes* biofilm formation on stainless steel and rubber and found that proteins that are negatively charged decreased the adhesion in comparison to the control. Hansson et al. [521] studied the adhesion of bacteria to diary packing materials and found that adhesion is dependent on the product contact time and surface. The authors emphasized that 10% of all fermented milk products remain on the inside of packaging material. In another study Uraz and Gündüz [522] analyzed fifty samples of raw milk and found that 44% of the sample was contaminated with *B. subtilis* and 4% with *B. cereus*.

Biofilm formation in the dairy industry has a significant impact on safety and quality. Dairy products are highly susceptible to contamination by biofilms, and it is challenging to eliminate those microorganisms. [581]

SEAFOOD-PROCESSING INDUSTRY

Like in the dairy produce industry, in the seafood-processing industry water quality and the equipment sanitation also have vital impacts on biofilm formation. [5, 582, 583] The common bacteria found in this industry are *V. cholerae*, *V. parahaemolyticus*, *V. vulnificus, E.coli, P. aeruginosa*, etc. [584-586] *L. monocytogenes, Salmonella* spp., *Bacillus* spp., *Aeromonas* spp., and *Pseudomonas* spp. [583] One reason can be low water quality due to cells detaching from the biofilm, resulting in the ineffectiveness of residual chlorine. [514, 587]

Guðbjörnsdóttir et al. [588] reported that adhered bacteria cells were found in the seafood-processing industry even if the sanitation and disinfection measures were provided correctly. From the shrimp processing plant, bacteria including *Pseudomonas* spp., *Aeromonas* spp., *Enterobacteriaceae*, and yeast were isolated. [589] In another study, Guðbjörnsdóttir et al. [588] demonstrated that *Pseudomonas* spp. promotes the adhesion of *L. monocytogenes* on stainless steel. Vestby et al. [590] found that *S. agona* and *S. montevideo* produce biofilms on fish products while *S. typhimurium* does not. Another conclusion of this study is that the persistence of *Salmonella* spp. in the factory environment depends significantly on the strain's biofilm-forming ability. Arkoudelos et al. [591] studied the ripening time for salter sardines and found no *S. aureus* or *S. enteritidis* after ninety days. Sushi, as a ready-to-eat food, can represent a potential source of pathogen bacteria. Atanassova and Klein [592] found that the prevalence of *E.coli* and *P. aeruginosa* was higher for fresh samples in comparison to frozen ones. They also reported that the quality of freshly prepared sushi strongly depends on the skills and habits of the cooks. In another study, Guerin [593] reported on an outbreak of food poisoning with contaminated processed fish products. Based on the questionnaires and retrospective studies, the reason for the poisoning was *S. livingstone* in a processed fish product.

MEAT-PROCESSING INDUSTRY

Meat is considered to be the food at greatest risk food of poisonous contamination. [10, 50, 524, 594] Dourou et al. [402] demonstrated that *E. coli* can attach itself to the surfaces of beef at various temperatures. Furthermore, Sala et al. [595] emphasized that several Gram-negative biofilms from beef surfaces are resistant to common antibiotics. Beef production facilities are often contaminated with Staphylococcus, Bacillus and Enterobacteriacae. [596]

From among different types of meat, poultry is considered to the riskiest type of meat. It has been shown that *Salmonella* spp. contamination can be related to dust, feces, poultry feed, and transportation. [597, 598] More importantly, more than half of the strains isolated on poultry farms were able to produce biofilms [597] The research of Chia et al. [599] demonstrated that *S. typhimurium* tends to adhere more efficiency to stainless steel than to rubber. Abdallah et al. [600] reported that *Salmonella* spp. isolated from Tunisian poultry produced moderate biofilms on PVC more than 50% of the time. In another study, Er et al. [601] proved that *S. typhimurium* biofilm is significantly more resistant to preservatives than planktonic cells are. In addition to *Salmonella* spp., *Campylobacter* spp. are also commonly found pathogens in poultry and poultry processing [547].

Several researchers have reported that temperature is an important factor of *C. jejuni* biofilm formation in the poultry industry. [498, 602] However, Hanning et al. [603] reported that *C. jejuni* can survive longer in biofilms at 32 °C, in comparison to the survival of planktonic cells and biofilms at 10 °C.

Van Houdt and Michiels [604] pointed out that the cells attached to the surface increased over time even under low temperatures, which could be a result of the migration of cells toward the surface by flagellar-mediated motility and Brownian motion. Several authors have proposed that bacterial cells can adhere and survive on food contact surfaces at low temperatures, and even increase in population as time progresses. [402, 605]

Another important factor of *E. coli* biofilm formation is the presence of other bacteria on the surfaces that can enhance or prevent biofilm growth. [596, 606, 607] Simpson Beauchamp et al. [608] evaluated different sanitizers in the meat industry and reported that acidified sodium chlorite and peroxyacetic acid-based sanitizers are the most effective. The study of Habimana et al. [609] showed that *E. coli* cells were embedded and covered by an *A. calcoaceticus* biofilm. Multispecies biofilms may increase the opportunities for pathogens to produce biofilms. [609, 610] Furthermore, it has been shown that *E. coli* cannot form a biofilm under high fluid flow due to shear stress. [611]

Ready-to-Eat Food Industry

The ready-to-eat food industry has increased in the last decade due to modern ways of food consumption. Since ready-to-eat food is not undergoing any thermal processes, this is the riskiest group of food being contaminated with microorganisms. [612]

Garrido et al. [613] studied the persistence of *L. monocytogenes* in refrigerators and found that smoked fish as ready-to-eat food was the most contaminated. Other researchers also found that the fish are the most significant sources of *L. monocytogenes.* [614] Podkowik et al. [615] found 96% staphylococcal isolates from seventy samples of ready-to-eat food. Among them, *S. aureus* and *S. epidermidis* were the most abundant. Moreover, they also reported that 80% of the isolated bacteria were resistant to at least four antibiotics. Nyenje et al. [616] studied *L. ivanovii* biofilm formation on ready-to-eat food in South Africa and found that *L. ivanovii* can form single and multispecies biofilms. This is of great concern regarding food safety as these organisms may adhere to cookware and cause cross-contamination. Keskinen et al. [617] studied *Listeria* spp. biofilms on stainless steel knifes for ready-to-eat food and found significantly greater transfer of strongly adhered biofilm to turkey than weakly adhered biofilm. Somers and Le Wong [618] found that *L. monocytogenes* biofilm was the most resistant on brick and conveyor materials in ready-to-eat production. Silagyi et al. [334] reported that *E.coli* is considered to be very important in ready-to-eat industry since it can form biofilms on food contact surfaces during food processing.

BIOFILMS ON FOODSTUFFS

Biofilms on foodstuff surfaces are different from those on food contact materials due to surface characteristics, nutrient accessibility, pH and other environmental characteristics. [619]

Moris et al. [475] showed that biofilms on endives and parsley represent about half of the total bacterial population on those plants. Multispecies biofilms were found in vegetables, such as spinach, lettuce, celery, and parsley, for which colonization of bacterial cells was proven in depth of 20 μm. [620] Similarly, Fett and Cooke [621] also found biofilms on sunflower seeds, clover, broccoli and alfalfa. Rudi et al. [622] reported on *Enterobacteriaceae*, *Pseudomonas* spp., *Oxalobacter* spp., and lactic acid bacteria on ready-to-eat salads. Furthermore, other authors found bacterial populations on lettuce [554, 563], apples [623], spinach [624], and cabbage. [625, 626] Elhariry [555] studied *B. cereus* biofilm on cabbage and found more than 80% of the strains produced biofilms.

More importantly, several authors have found that bacterial populations tend to colonize the vegetable and fruit stomata, trichomes, veins, and cell wall junctions. [624, 627] For example, Warner et al. [628] found *S. thompson* in the stomata and veins of spinach leaves. In Brazil, more than five hundred samples of foodstuff were tested for *L. monocytogenes*, and 3% of the samples were positive. More importantly, most strains were resistant to 125 ppm of chlorine-based disinfectant [629], which indicates that while a low share of samples were positive, those that were positive were highly resistant.

In addition to naturally occurring irregularities on food surfaces, punctures, cuts, splits, and cracks due to injury during pre-harvesting and post-harvesting handling can also promote bacterial attachment and consequently biofilm formation. [623] Lu et al. [336] showed formation of biofilms on the damaged surfaces of apples and oranges. Biofilms were also detected on the cracks of the hypocotyls and cotyledon of alfalfa. [564] Similarly, Liao and Cooke [560] analyzed the *Salmonella* spp. biofilm formation on peppers and found significantly more bacteria on injured than on intact ones. Khalil and Frank observed that leafy green vegetables are an ideal niche for biofilm formation. [630]

During production from farm to fork, food can be contaminated with a wide range of microorganisms and ideal environments on the foods' surfaces can promote their development.

FOOD CONTACT MATERIALS AND BIOFILM FORMATION

Food contact materials have a significant impact on bacterial attachment and biofilm formation. The most commonly used materials in the food industry are stainless steel, glass, rubber, polyurethane, [329, 487, 599, 631] Teflon, nitrile butyl rubber [632] and wood. [633] Microorganisms are able to form biofilms on various materials. As reported by Stepanovic et al. [378], *Salmonella* spp. and *L. monocytogenes* form biofilms on plastic material. Since food represents an ideal environment for the growth of bacteria, even weak strains can produce biofilms that are difficult to eliminate. [634] Adetunji and Isola [539] analyzed biofilm formation on various materials, such as wood, stainless steel and glass, and found the highest biofilm formation on wood surfaces and the lowest on the glass; while stainless steel can prevent colonization, it is vulnerable to corrosion. *Salmonella* spp. can attach itself to different surfaces, but it seems that more cells are attached to Teflon surfaces and stainless steel and to polyurethane, which indicates that material properties have a great impact on initial adhesion. [599] The authors also stated that surface roughness has no impact on adhesion; many other authors have reported the opposite. [635, 636] Bernardes et al. [637] studied the adhesion of *B. cereus* on stainless steel and found temperature and time to be the most important factors promoting attachment.

In the study of Arnold et al. [638], fewer bacterial cells were attached to the electro-polished stainless steel surface than on untreated steel that was rougher. Guobjörnsdóttir et al. [588] pointed out that not only the material properties, but also the hygienic design such as welding, joints, corners, and equipment design, could have a significant impact on biofilm formation.

Stainless steel is a frequently used material in food production and handling. Regular sanitation can damage stainless steel surfaces. Organic material with microorganisms can gather in these damaged sites, forming biofilms not easily exposed to the antimicrobial agents. Boyd et al. [639] studied the different types of cleaning, e.g. spraying and brushing, and have found brushing to be more efficient than spraying. The retention of organic matter was greater on rough samples. [639]

Møretrø et al. [640] found that *Staphylococcus* spp. formed firmer biofilms in the presence of sodium chloride and nutrients. Important findings of Borucki et al. [641] showed that strains of *L. monocytogenes* that are frequently present form more abundant biofilms in dairy production that those that appear sporadically.

Midelet and Carpentier [642] showed how environmental conditions are important in biofilm formation, having proved that *P. fluorescens* and *S. sciuri* form EPS when calcium is added. The authors also found that the cells of *P. fluorescens* detached quickly when cleaned with a chlorinated alkaline agent, but the attachment strength of *S. sciuri* was increased. Treatment with disinfectant based on glutaraldehyde and a quaternary ammonia also increased the efficiency. [642]

Conclusion

Bacteria can form biofilms on all food contact materials, although the rate of adhesion strongly depends on the material's surface characteristics. [643, 644] Nevertheless, proper sanitation should prevent biofilm growth and maturation, which can assure the good hygiene practice that is the basis of safe products. Sharma et al. have recommended supervising the operating time between cleaning and sanitation to prevent mixed species biofilm formation. Lindsay [645] reported that cleaning with short intervals is effective in preventing biofilm formation. According to good hygiene practice, equipment will reduce the dirt formation and promote cleaning processes that can result in decreased bacterial attachment on food contact materials. In addition, the surfaces of the material that is already in use can be altered with an antifouling coating.

Chapter 7

BIOFILMS IN HOSPITAL FACILITIES

"If you don't think like bacteria, you're on the wrong planet"
(Craig Venter)

Medical devices are implemented into the human body to substitute for or restore biological functions. [646, 647] Understanding and treating clinically relevant infections is vital, especially those related to temporarily or permanently implanted devices. However, contact between the human body and artificial material can result not only in rejection, but also in device-related infections (Table 2). [54, 648] Since the inclusion of artificial materials, such as orthopedics, catheters, vascular stents, etc., can cause a foreign-body response, this will result in biofouling. [648] Moreover, invasive surgical procedures can increase medical-device-associated infections. Donlan and Costerton estimated that 65% of hospital-acquired infections are related to biofilm formation on medical devices. [12, 649] An interesting study by Henry et al. [650] demonstrated that even uninfected implants are colonized with bacteria. Moreover, cultivating non-infected implants showed a 70% positive rate and more than 90% were colonized with at least one species of bacteria. Similarly, Silverstein et al. [651] determined the presence of biofilm on 80% of non-infected penile prostheses.

There is some evidence that hospital-related illnesses, such as cystic fibrosis, endocarditis, otitis media and periodontitis, appear due to biofilm growth on medical devices or human tissue. [369] Nevertheless, in addition to devices, the water distribution system and air conditioning can be sources of *L. pneumophila*, mycobacteria, molds and yeasts. What is less clear is an understanding of how the interaction and growth of pathogenic organisms in a biofilm result in an infectious disease process. [652] Medical device biofilms are a critical issue of hospital hygiene, since they can represent a source of infection or even a cause of death. Particles that are detached from biofilms can travel along the bloodstream and cause blood clots, and can infect the urinary tract or other body systems. The resistance of hospital-relevant biofilm is also related to the plasmid exchange of bacteria. Bacterial cells in biofilms are much more resistant to antimicrobial agents than those in bulk liquid. Nevertheless, bacteria in biofilms can produce toxins, causing intoxication. Above all, biofilms are resistant to human immune system responses and can be lethal for patients with immune diseases. [12] In hospital facilities common materials, such as stainless steel, titanium, and polycarbonate, are used. Many studies report that *S. epidermidis*, *S. aureus* and *E.coli* are frequently found on medical devices (Table 2).

In the case of invasive procedures, the medical device is in direct contact with human body tissue, which can result in infection. Hospital-related infections are an important public health issue that also has economic significance. [677, 678] McGee and Gould [679] reported that in the USA more than 80,000 cases of bacteremia were related to catheters. Meanwhile, in Europe more than 1000 deaths per year are reported as a result of catheter-related infection. [680]

Biofilms in hospital facilities have a significant impact on human health, especially taking into the account the fact that people in this environment are more susceptible to infection. Biofilms are becoming increasingly resistant to antimicrobial treatment and to the immune systems of hosts. Therefore, new strategies for combating medical device biofouling have to be implemented. [12] One crucial step is to prevent the colonization of bacteria on hospital surfaces. With reduced bacterial adhesion, the risk of biofilm formation could be reduced or delayed. [681] A biomaterial that prevents bacterial adhesion depends mainly on the location, time of use, level of acceptable contamination and the probability of infection. If the medical device does not penetrate the human body and is not in direct contact with blood and other body fluids, it does not represent high risk of infection. However, if the material is in direct contact, then the possibility of device-related infections increases. [54]

One of the most important medical devices that come in direct contact with the human body are those that are in contact with mucous membranes, such as the bronchial mucosa, urothelium, endometrium and conjunctiva. As reported by Rello et al. [682], intubated patients are at a high risk for ventilation pneumonia, which often causes morbidity. It has been reported that endotracheal intubation increases the risk of pneumonia 20 fold as the endotracheal tubes enable the migration of bacteria cells from the contaminated oropharyngeal space to bronchial mucosa. [683] Several authors have reported that infection caused by urinary catheters is estimated to be about 3-10%. [684] The high prevalence of medical-device-associated infections in medical facilities has caused an increase in the use of antimicrobial therapy, which has consequences in the multidrug resistance of biofilms. [685, 686]

Table 2. Common bacteria forming biofilms on medical devices

Biofilm type	Contact materials	References
S. epidermidis	stainless steel, glass, titanium	Chaieb et al. [653] Cerca et al. [654] Braceras et al. [655] Braem et al.[656] Van der Borden et al. [657]
S.aureus	ceramic, stainless steel, titanium	Batalu et al. [658] Karu et al. [659] Braceras et al. [655] Kiedrowski et al. [660] Guembe et al. [661]
E.coli	titanium, polymers	Tian et al. [662] Pradeep Kumar et al. [663] Shah et al. [664] Shi et al. [665] Hu et al. [666]
C. albicans	polycarbonate, Teflon, stainless steel	Shinde et al. [667] Estivill et al. [668] Mishra et al. [669] Raman et al. [670] Chassot et al. [671]
P. aeruginosa	polyvinyl chloride, titania	Tarquinio et al. [672] Loo et al. [673] Darouiche et al. [674] Burnham et al. [675] Roosjen et al. [676]

Contact Lens Biofilms

One of the most commonly used medical devices are contact lenses; millions of people use them every day. The ocular surface is unsterile due to contamination from surrounding environment. *Staphylococcus* spp. and *Corynebacteria* can commonly be found on the conjunctiva. [687, 688] The normal ocular flora and other defense mechanisms can prevent infection in a normal and healthy eye. [689] Tear fluid and the corneal epithelium are the first line of defense against ocular infection. Naturally occurring substances in the cornea, such as β-defensins, cathelicidin, and cytokeratin-derived antimicrobial peptides, prevent infection. [690] Since contact lenses are foreign bodies, they can alter the homeostasis by changing the normal flora, which can result in inflammation. [691] There are several infectious illnesses related with the use of contact lenses, including acute eye occulitis, keratitis, infiltrative keratitis and peripheral ulcer. [687, 692, 693] The adhesion of bacteria to lenses deepens on the species and the physical and chemical properties of contact lenses material. Bacteria tend to adhere more frequently to hydroxyethylmethacrylate-based lenses in comparison to silicone hydrogel lenses due to their water content. [692] Dutta et al. [692] reported that *P. aeruginosa*, *S.marcescens* and *S.aureus* are frequent causes of keratitis. Shousha et al. [694] studied keratitis on rat contact models and found clinical signs of biofilm infection, such as increased corneal thickening and loss of epithelial continuity.

Interesting results were reported by Szczotka-Flynn et al. [687]: in 60% of normal healthy users of contact lenses the surface was covered with pathogenic bacteria. Stapleton et al. [695] reported that new materials, including hydrogel or silicon hydrogel, do not efficiently prevent keratitis. In another study, Burnham et al. [675] concluded that chronical corneal inflammation enhances *P. aeruginosa* biofilm formation on silicone hydrogel contact lenses. Kaynak Onurdağ et al. [696] studied biofilm formation on different contact lens materials and reported the low adhesion of *P. aeruginosa* on the materials polymacon, etafilcon A, hilafilcon, ocufilcon and lotrafilcon. The levels of adherent yeast were higher in lotrafilcon and ocufilcon materials. Some authors report that biofilms are present in eye implants and other prostheses in ophthalmology. [697, 698]

Intrauterine Contraceptive Device Biofilms

It must be understood that every material that is incorporated into the body acts as a foreign body and affects the normal immune response. One example is intrauterine contraceptive devices, which can cause infection.

Some studies have reported that use of intrauterine contraceptive devices can cause cervical infection caused by *E. coli, Trichomonas vaginalis* and *Candida* spp. [699, 700] Abdel-Hafeez et al. [701] found the almost 50% of intrauterine contraceptive devices are colonized by biofilms. Other authors emphasize that the use of intrauterine contraceptive devices based on copper does not increase the risk of contracting HIV, chlamydial or gonococcal infections. [702] Precipitation of the metal in the form of copper has been shown to act as an antibacterial agent. [703] However, it must be understood that precipitation is often a bacterial mechanism of resistance to toxic heavy metals. Actinomycosis has also been reported in association with plastic devices, and some recent studies show that some additional antibacterial substances should be added to intrauterine contraceptive devices. [704] Carrillo et al. [705] found *A. israelii* biofilm on copper-based intrauterine contraceptive devices. Chassot et al. [671] found strong adherence potential on copper-based intrauterine contraceptive devices and concluded that they represent a potential *C. albicans* reservoir. An interesting case of intrauterine contraceptive device infection by *Streptococcus agalactiae* was reported by Ducharne et al. [706] as they found the same bacteria on intrauterine contraceptive device as on hip prostheses. Farley et al. [707] conducted research on infection risk related to the intrauterine contraceptive devices and found that infection was present only in 0.4% of inserted intrauterine contraceptive devices. However, other studies report that the long-term use of intrauterine contraceptive devices can result in actinomycosis. [708, 709]

Long-term gynecologic infections can cause cancer of the ovaries, fallopian ducts or uterus. [703, 709]

Partially Implanted Device Biofilms

The skin and mucosal membranes are natural barriers for microorganisms, but most of them cannot overcome these barriers. Percutaneous and permucosal implants break this barrier and provide direct entrance of bacteria into the human body. This situation can cause the spread of the infection to the other tissues, organs and consequently result in bacteraemia or septicaemia.

[710] If external fixation systems are applied to reduce traumatic fractures, infections can develop from wound contamination. However, for percutaneous implants, initial wound contamination is not necessary. The points where percutaneous pins and wires penetrate the tissues, breaching the skin barrier, also termed "pin sites," are particularly exposed to the risk of colonization and infection. [678] Hargreaves et al. [711] examined the formation of biofilms on percutaneous wires for fracture fixation and found that 25% of the samples were positive for *Staphylococcus* spp.

Some studies indicate a device-related infection rate of 50% for femoral shaft fractures, followed by 35% for diaphyseal fractures of the tibia, 20% for humeral shaft fractures, 12.5% for diaphyseal fractures of the forearm, and less than 5% for tibial plateau fractures. [712-714] After surgical implants, the care of the wound should be considered to be the most important factor for the prevention of infection, including proper wound care at the pin site, biomaterial surfaces with good tissue integration, and bioactive biomaterials endowed with antibacterial properties. [715-717] Tillander et al. [718] reported on infection related to the osseo-integrated implants for limb prosthesis fixation and showed that 40% of patients have some kind of infection, but only less than 3% resulted in implant failure.

Implants, such as vein chatters, ventricular assist devices, or other devices in contact with the circulatory system, can cause infection of the cardiovascular system or central nervous system. [719-723] Dialysis-related infections are significant as they represent nearly a quarter of all infections among dialysis patients and are related to high morbidity. [723, 724] As shown in Figure 15, a vein catheter use can result in bacteremia due to migration of the skin flora from the surface close to the exit site (Figure 15a); colonization of the catheter interior surfaces (Figure 15b) or hematogenous dissemination (Figure 15 c). [725]

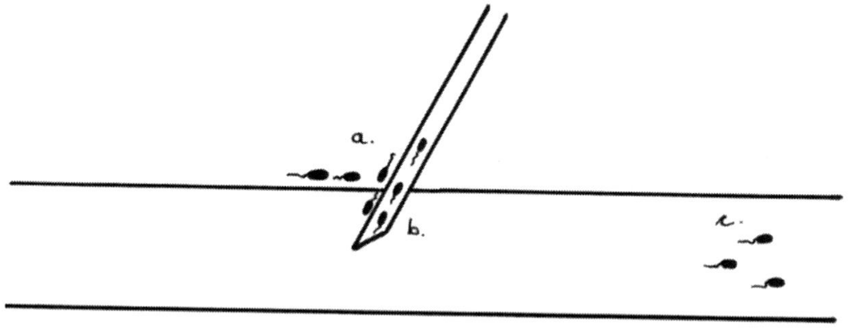

Figure 15. Biofilm growth due to vein catheter use.

surfaces will prevent the colonization of bacteria on tooth . Therefore, antibacterial coating strategies can be ent infection. [744]

INTERNAL IMPLANT DEVICE BIOFILMS

implant devices are implanted medical devices and are by tissues. The infections due to totally internal implants ause the surface contamination before or during the ss or by hematogenous seeding from a distant place.

seeding can represent a secondary source of infections for ecially those in direct contact with the blood stream and the tem. Charlier et al. [747] reported on *L. monocytogenes*-nd joint infections and found a 15% infection rate among hors also concluded that osteoarticular listeriosis affects and occurs in immunocompromised patients. The illness antibiotic treatment with implant removal or replacement for tudy, Monksfield et al. [748] examined sixteen used internal and found biofilm colonization deep in the threads on eight. . [749] study *S. aureus* biofilm on different materials for nts and found minimal differences among biofilms formed on ate and titanium.

eported that 43% of the patients with nosocomial bacteraemia tic valve endocarditis. [750] Chamis et al. [751] found that of cardiac implants were contaminated with *S. aureus*. The hematological seeding with *S.aureus* was found for orthopedic It has been proven that *S. aureus* was present on the surface of devices in the first month after surgery, and that after that *ative staphylococci* were found. [753-755] Although the for cardiac implant devices is low, more that 1.3 million implantable electronic devices are implanted globally each ail et al. [757] reported that the infection rate for pacemakers is and 20% and up to 0.8% for cardiovascular defibrillators. In Gandhi et al. [758] reported that the mortality at one year after rdiovascular implants is between 8% and 40%.

Attenello et al. [726] i
of implants in the head i
addition to implant fail
psychomotor retardation anc
measures for percutaneous a
of devices. [724]

DENTAI

Biofilms forming on dent
caries and periodontal diseas
permucosal implant. [728] It h
implants is increasing every yea
implant diseases. Zitzmann and
related to dental implants: peri-i
implant mucositis is an infectic
presenting with exudates, swelling
implant pocket formation and the
implants. [731] Dental plaque co
[732, 733], such as *P. aeruginos*
epidermidis and others. [734] Ceram
implants and are known to be low
characteristics. [299] Lima et al. [73
ceramics. Moreover, ceramic is more
[736] On amalgam material, Auschil
biofilms in comparison to only a 6-
researchers have shown that resin com
reason for this could be the release c
triethyleneglycol dimethacrylate from
interesting study by Ready et al. [741] sl
the presence of amalgam and could evei
[742] reported that 30% of patients tre
some complications in the first five year
implants has been estimated to be 7% at 1
showed that contamination of dental imp
process of implantation, and that mucosal
plaque or supra gingival parts.

Smooth material
or implant surface
implemented to prev

TOTALLY

Totally interna
totally surrounded
occur mainly bec
implantation proc
[745, 746]

Hematogenou
some implants, es
cardiovascular sy
associated bone a
patients. The au
prosthetic joints
requires intensive
cure. In another
screws with SEM
Lauderdale et a
orthopedic impla
glass, polycarbo

It has been
develop prosthe
about one third
highest rate for
implants. [752]
cardiac implan
Coagulase-neg
infection rate
cardiovascular
year. [756] So
between 0.13
another study
surgery for ca

Attenello et al. [726] reported hydrocephalus as one of the consequences of implants in the head in approximately 15% of cases. Nevertheless, in addition to implant failure, possible consequences are lowered IQ, psychomotor retardation and higher risk of seizures. One of the best preventive measures for percutaneous and permucosal device biofilms is the rational use of devices. [724]

Dental Implant Biofilms

Biofilms forming on dental hard or soft tissue represent a key cause of caries and periodontal disease. [727] Dental implants are a type of the permucosal implant. [728] It has been reported that the number of dental implants is increasing every year, [729] which also increases the risk of peri-implant diseases. Zitzmann and Berglundh [730] distinguished two illnesses related to dental implants: peri-implant mucositis and peri-implantitis. Peri-implant mucositis is an infection involving soft gingival tissues, usually presenting with exudates, swelling and bleeding. Peri-implantitis includes peri-implant pocket formation and the loss of supporting the bone around dental implants. [731] Dental plaque consists of several different bacteria species [732, 733], such as *P. aeruginosa*, *Enterobacteriaceae* ssp., *S. aureus, S. epidermidis* and others. [734] Ceramic materials are commonly used for dental implants and are known to be low adhesive materials due to their surface characteristics. [299] Lima et al. [735] found low plaque adhesion on zirconia ceramics. Moreover, ceramic is more hygienic than the natural tooth surface. [736] On amalgam material, Auschill et al. [737] found up to 17-μm thick biofilms in comparison to only a 6-μm layer on ceramic material. Several researchers have shown that resin composites enhance bacterial adhesion. One reason for this could be the release of ethyleneglycol dimethylacrylate and triethyleneglycol dimethacrylate from composite resins. [299, 738-740] An interesting study by Ready et al. [741] showed that some species could grow in the presence of amalgam and could even assimilate mercury. Pjetursson et al. [742] reported that 30% of patients treated with dental implants will have some complications in the first five years of insertion. Total failure of dental implants has been estimated to be 7% at 10 years. [743] Norowski et al. [728] showed that contamination of dental implants can be introduced during the process of implantation, and that mucosal surfaces can be infected by dental plaque or supra gingival parts.

Smooth material surfaces will prevent the colonization of bacteria on tooth or implant surfaces. Therefore, antibacterial coating strategies can be implemented to prevent infection. [744]

Totally Internal Implant Device Biofilms

Totally internal implant devices are implanted medical devices and are totally surrounded by tissues. The infections due to totally internal implants occur mainly because the surface contamination before or during the implantation process or by hematogenous seeding from a distant place. [745, 746]

Hematogenous seeding can represent a secondary source of infections for some implants, especially those in direct contact with the blood stream and the cardiovascular system. Charlier et al. [747] reported on *L. monocytogenes*-associated bone and joint infections and found a 15% infection rate among patients. The authors also concluded that osteoarticular listeriosis affects prosthetic joints and occurs in immunocompromised patients. The illness requires intensive antibiotic treatment with implant removal or replacement for cure. In another study, Monksfield et al. [748] examined sixteen used internal screws with SEM and found biofilm colonization deep in the threads on eight. Lauderdale et al. [749] study *S. aureus* biofilm on different materials for orthopedic implants and found minimal differences among biofilms formed on glass, polycarbonate and titanium.

It has been reported that 43% of the patients with nosocomial bacteraemia develop prosthetic valve endocarditis. [750] Chamis et al. [751] found that about one third of cardiac implants were contaminated with *S. aureus*. The highest rate for hematological seeding with *S.aureus* was found for orthopedic implants. [752] It has been proven that *S. aureus* was present on the surface of cardiac implant devices in the first month after surgery, and that after that *Coagulase-negative staphylococci* were found. [753-755] Although the infection rate for cardiac implant devices is low, more that 1.3 million cardiovascular implantable electronic devices are implanted globally each year. [756] Sohail et al. [757] reported that the infection rate for pacemakers is between 0.13% and 20% and up to 0.8% for cardiovascular defibrillators. In another study, Gandhi et al. [758] reported that the mortality at one year after surgery for cardiovascular implants is between 8% and 40%.

CONCLUSION

In recent decades, medical implantations have increased exponentially with the aims of curing and restoring the human body and of increasing the quality of human life. However, at the same time, new issues have appeared regarding device-related biofilms and the consequences of infection. [759] Conventional antibiotics are efficient against planktonic cells, but this cannot be taken granted for bacteria in biofilms. Solutions include materials with properties to prevent bacterial adhesion or with incorporated antibacterial agents. The best strategy to reduce the infection risk is to achieve hygienic conditions before and during the implantation operation, assuring proper antibiotic prophylaxis, using bioactive materials and caring for post-operative wounds. [760, 761]

Chapter 8

PREVENTION OF THE GROWTH OF BIOFILMS

"An ounce of prevention is worth a pound of cure"
(Ben Franklin)

The prevention of biofilm growth on food contact material or materials in hospital facilities is related to the chemical and physical properties of the materials. [762, 763] Antimicrobial surfaces can prevent adhesion through passive or active approaches. The surface properties can be modified by changing the roughness and hydrophobicity, and applying the functionalization of the surface with peptides that disable the bacteria and prevent adhesion and, consequently, biofilm growth. Active approaches include drug-eluting materials to disable pathogens and inhibit adhesion. [648]

PASSIVE STRATEGIES FOR THE PREVENTION OF BIOFOULING

Hydrophilic Surfaces

Hydrophilic surfaces prevent cell adhesion to the material with the formation of a physical barrier known as a "hydration layer" (Figure 16). [764, 765] The bonding between functional groups on the surface and water forms such a layer. [765]

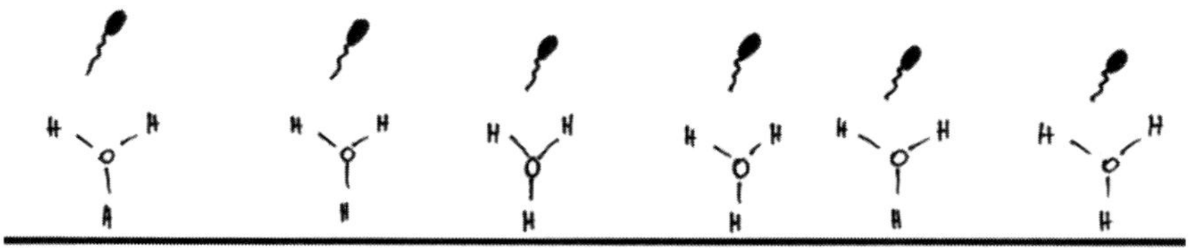

Figure 16. Hydrophilic surface preventing biofilm formation.

The most commonly used material for medical devices is polyethylene glycol [766], which can resist protein adsorption and cell adhesion due to the formation of a hydration layer. In addition to polyethylene glycol, some naturally occurring polysaccharides, such as chitosan and dextran can prevent bacterial adhesion. [767] Some researchers have reported that heparin or hyaluronic acid may also act as hydrophilic coatings. [768] One example in practice is heparin-coated urinary catheters, which can reduce biofilm formation. [769-771] Gao et al. [772] analyzed different polymers in relation to biofilm prevention and found low biofouling on polymer-coated materials. In contrast, Bonsaglia et al. [556] found that *L. monocytogenes* tends to adhere more intensively to hydrophilic surfaces than to hydrophobic ones. Andersen et al. [773] reported that hydrophilic coated silicone promotes *E. coli* adhesion. Results of another study showed that a highly hydrophilic, photo-reactive coating of TiO_2 nanoparticles was highly effective for the prevention of biofilm formation on ceramic ultrafiltration membranes. [774] With plasma treatments, a hydrophilic surface with antifouling and antimicrobial activity can be created. [775, 776] Since bacteria are negatively charged at pH 7, it is obvious that cells adhere better to positively charged materials and less so to negatively charged surfaces. [777, 778] However, long-term exposure of hydrophilic materials in biological systems results in the destruction of the hydration layer, due to surface oxidation. [765]

Hydrophobic Surfaces

The other mechanism for the prevention of bacterial adhesion and the consequent formation of the biofilms is hydrophobic surfaces, which repel the attachment of water molecules. In order to create hydrophobic properties of material, functional groups that resist hydrogen bonding onto the surface of the material have to be implemented (Figure 17).

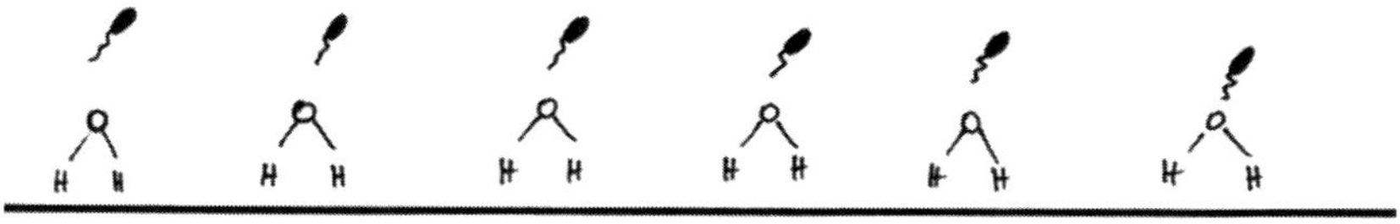

Figure 17. Hydrophobic surface preventing biofilm formation.

Some examples of hydrophobic surfaces include fluorinated coatings, silica colloids and nanoparticles. [779-781] Biodegradable hydrogels derived from Vitamin E show some potential to prevent biofouling due to their hydrophilic character. [782] Furthermore, surfactants on a material's surface can prevent bacterial adhesion by increasing the hydrophilicity of the surface. [765, 783] Zeraik and Nitschke [784] showed that surfactant coatings can significantly decrease bacterial attachment. Brambilla et al. [785] demonstrated that highly hydrophobic dentin prevents the formation of *S. mutants* biofilms in an *in vitro* model. Zhong et al. [786] examined nafion-coated stainless steel materials and found very low adherence of *E.coli* due to the hydrophobic properties of material's surface. Meanwhile, Vani et al. [787] emphasized that reducing the surface hydrophobicity alone is sufficient to develop a biocompatible material, but providing favorable surface functional groups was also a necessary criterion. In a study by Rzhepishevska et al. [117], hydrophobic polymethylmethacrylate was found to be efficient in preventing bacterial adhesion and biofilm formation.

Zwitterion Charged Coatings

Zwitterionic antifouling materials are used to prevent bacterial cell adhesion by forming both positively and negatively charged species on the surface of the material and resisting the nonspecific adsorption of proteins (Figure 18). [788]

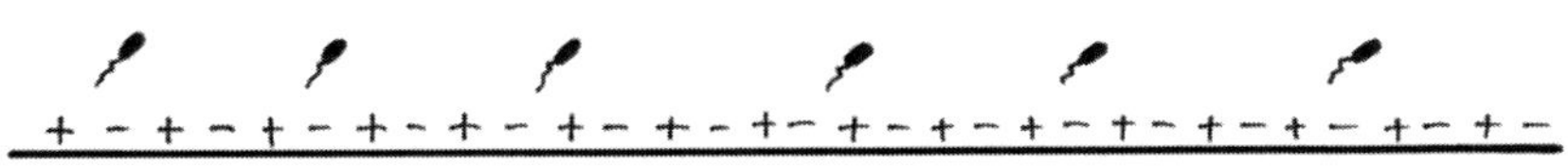

Figure 18. Zwitterionic surface preventing biofilm formation.

For example, polysulfobetaine and polycarboxybetaine are used to form zwitterionic surfaces. [648, 789] In comparison to the hydrophilic surfaces for which the hydration layer is formed by weak hydrogen bonds, with zwitterionic surfaces, the hydration layer is tightly formed via electrostatic interactions. [648] Since zwitterionic surfaces possess an equally distributed charge along the entire surface, adhesion and consequently biofilm production are prevented. [790, 791] Li et al. [792] tested different coating containing zwitterion function groups, such as amino, hydroxyl, and methyl monolayers, and found that coated surfaces are highly resistant to bacterial adhesion and biofilm formation. Chang et al. [793] reported poly(vinylidene fluoride) with zwitterionic functional groups to be the best antifouling, anticoagulant, and antihemolytic material for medical devices. The incorporation of zwitterions suppressed the adsorption of extracellular polymer substances and the adhesion of bacteria. [794]

Cationic Charged Coatings

Surfaces with positive charges have antimicrobial activity and can prevent biofouling (Figure 19). [795, 796] The positive charge destroys the lipid membrane of bacteria and causes cell lysis.

One example of cationic charge coating is chithosan which, has antimicrobial effects on bacteria, fungi and viruses, and has low toxicity for human cells. [797, 798] Smith et al. [799] reported that cationically charged dicocodimethylammonium chloride prevents the formation of biofilms; after twelve weeks, only low-level biofouling was detected. Williams et al. [800] studied polydimethylsiloxane coating in an animal model and found one hundred percent biofilm prevention, while a biofilm developed on a control material. Ding et al. [801] examined intravascular catheters with cationic polycarbonate and found anti-fouling action against *S. aureus* biofilm.

Nevertheless, some authors reported that wide use of cationic charge coatings can lead to the bacterial resistance. [802, 803]

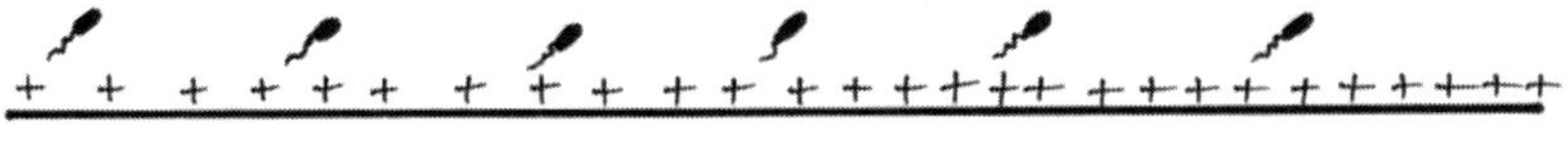

Figure 19. Cationic surface preventing biofilm formation.

Biomolecule Coatings

It is believed that biomolecules that are deposited on the surface can provide a barrier between cells and the material surface. [474] Biomolecules are attached to the surface by covalent attachment and have non-specific antimicrobial effects. [804-806] The peptides are mainly derived from natural sources with the advantage of being biocompatible; because they form a deposition layer, biofouling by other molecules is prevented (Figure 20). [805, 807-809]

Waterhouse et al. [804] reported that antimicrobial peptides inhibit bacteria, such as *P. aeruginosa* and *S. aureus*. Moreover, they found that metal stents coated with antimicrobial peptides reduce the formation of thrombus. Since device rejection is at the greatest risk immediately after surgery, peptide coatings could also reduce device rejection. [760, 810] Raman et al. [670] reported on a polymer with β-peptides that eliminated *C. albicans* planktonic cells and biofilm formation on catheters. Holmberg et al. [811] studied a titanium-coated dental implant with an antimicrobial peptide, GL13K, derived from parotid secretory protein and found significant reductions in biofilm formation. Similar findings were reported by Kazemzadeh-Narbat et al. when studying titanium surfaces with antimicrobial peptides against *S.* aureus and *P.* aeruginosa. [812] Forbes et al. [808] demonstrated that human apoliprotein E is an effective antiseptic coating agent. Moreover, Shukla et al. [813] reported that ponericin G1, an antimicrobial peptide, is effective against biofilm.

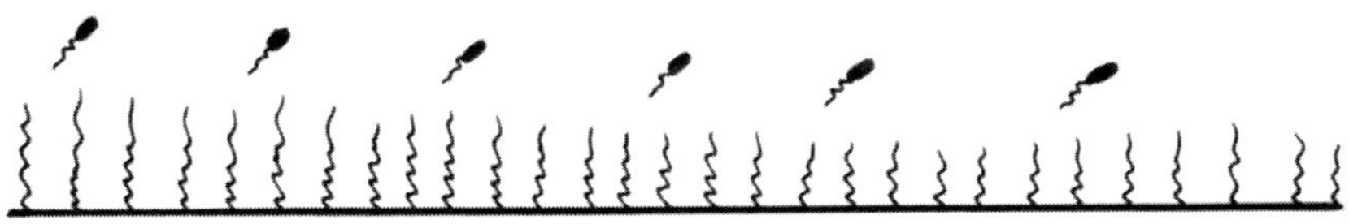

Figure 20. Biomoleculescoatings preventing biofilm formation.

Surface Roughness

Materials with low roughness prevent biofilm formation since the contact surface for adhesion in sub-micro scale material is decreased due to the small gaps for the presence of bacteria (Figure 21). In contrast, at greater surface roughness, a greater specific area creates more available active sites for thermodynamic reactions. [72]

Figure 21. Low roughness surfaces preventing biofilm formation.

Several studies indicate that increasing the roughness results in promoted adhesion. [73, 168, 681] The threshold for bacterial adhesion can be a surface roughness of Ra = 0.2 μm, below which no further reduction in bacterial adhesion could be expected. The increase in roughness above this threshold results in a simultaneous increase in cell deposition. [167, 168, 814, 815] In contrast, some studies reported that roughness has no impact on the adhesion of *S. epidermidis* and *P. aeruginosa.* [816] In the literature, the findings on roughness are contradictory. Some researchers report a positive relation between bacterial adhesion and increased roughness [72, 817, 818], while other researchers state the opposite, i.e. that there is no relation. One explanation could be the use of different definitions of roughness, bacteria strains and environmental conditions in the adhesion tests.

Surface Topography

In addition to the roughness, the surface topography also governs bacterial adhesion (Figure 22). Irregular surface properties can promote cell adhesion as bacteria tend to adhere in gaps and cracks. [681] Although fewer studies have been made on surface topography than on roughness, some of them indicate that adhesion is also increased with increased surface irregularities. [819-821]

Figure 22. Material irregularities promote biofilm formation.

It is necessary to determine which parameters of nanofeatures are most effective in reducing the adhesion of bacteria. Some research has shown that topographic or nano-roughness modifications can change surface properties such as wettability. [72, 822] The surface of titanium nanotubes is more hydrophilic than a conventional titanium surface. [823, 824] Changing the surface properties has applications in the management of bacterial adhesion

[825, 826]. Moreover, nanotubes can consist of biocides to improve their action against biofilms. [826-830] Nano-surfaces can have different shapes: nanotubes, grooves, channels or ridges, pits or pillars. [823, 831, 832] Regarding the spatial organization of nanofeatures, the distance between them can have a significant impact on cell adhesion. Hochbaum and Aizenberg [821] reported that *P. aeruginosa* increases adhesion by increasing the distance among adjacent posts.

The orientation of nanostructures can form patterns that have an impact on microbial adhesion. The positioning of the ridges on polymers with these types of nanofeatures can enhance or inhibit bacterial adhesion. [820] Another study by Chung et al. [831] reported that the lack of adhesion of *S. aureus* on polydimethyl siloxane elastomer was bio-inspired from shark skin.

ACTIVE STRATEGIES FOR THE PREVENTION OF BIOFOULING

Evidence exists that the non-critical use of antibiotics results in bacterial resistance. [833, 834] One solution can be a local application on the material surface. [835] The challenge of active materials is done to achieve efficient control on bacteria and to simultaneously assure biocompatibility with surrounding tissues.

Several authors have demonstrated incorporated antibacterial substances on surface materials that can prevent bacterial and thrombus formation. [836-838] In addition to antibacterial agents, immunosuppressants, and antiproliferative drugs can also be incorporated into device matrices to increase the possibility of successful implantation. [839-841]

Surface Modifications

Surface modifications can substantially reduce the adhesion of bacterial cells. Nowadays, several different coating are available to prevent bacterial adhesion. Polymer brush coatings are known for their antibacterial properties. A polymer brush is formed by hydrophilic chains that adhere to a surface in high density, forcing the polymer chains to stretch from the surface into the bulk solution (Figure 23). [842]

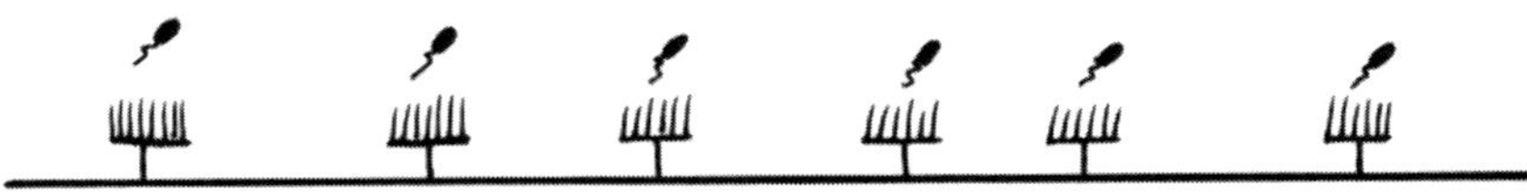

Figure 23. Brush-coated surfaces prevent biofilm formation.

Tiller et al. [843] modified the surface of glass with poly(4-vinyl-N-hexylpyridinium bromide) and demonstrated only 5% adhesion of *S. epidermidis, P. aeruginosa*, and *E. coli*. Similarly, Roosjen et al. [844] inhibited bacteria and yeast with polyethylene oxide by decreasing the Lifshitz-van der Waals attraction between the cells and the glass surface. In another study by Roosjen et al. [845], ethylene oxide brush-coated glass reduced adhesion by 98%. In contrast, some studies reported that *P. aeruginosa* produces surfactants that adhere to the poly-ethylene oxide due to acid-base interactions. [846, 847] Furthermore, polyethylene glycol coatings have been shown to have antibacterial properties. [846, 848]

The use of radiation-induced graft polymerization can introduce long graft chains on common polymer materials [817], and can also modify different polymers, which may prevent protein and bacterial adhesion. Gottenbos et al. [777] analyzed bacterial cell adhesion on positively charged surfaces and demonstrated that positive charges can promote adhesion, but biofilm formation is slower. Terada et al. [849] reported that 80% of adhered *E. coli* cells and 60% of *B. subtilis* cells on positively charged surfaces were inactivated after a contact time of eight hours.

Monomeric silicon chemicals have low surface energy and are highly hydrophobic [850, 851]. Organo-silane with a quaternary amine group acts in an antibacterial manner and can prevent biofilm formation [852]. Quaternary ammonium salts have antibacterial properties, and they are often used as disinfectants in the food industry. They also act against fungi, protozoa and viruses. The main mechanisms of quaternary ammonium salts is to disrupt cell membranes. [853] Gottenbos et al. [854] tested the antibacterial properties of an organo-silane quaternary ammonium compound, 3-(trimethoxysilyl)-propyldimethyloctadecylammonium chloride, coated on silicone rubber, and reported substantial decrease in adhesion. Coating with quaternary ammonium salts was also investigated for its efficacy against mixed fungal and bacterial biofilms by Oosterhof et al. [855]. Gong et al. [856] provided coating with partially hydrolyzed quaternary ammonium salts co-polymerized with 2,2-[4(2-hydroxy 3-methacryloxypropoxy)-phenyl] propane and demonstrated prevention of the adhesion of *S. mutans*, *A. naeslundii* and *C. albicans*.

Kregiel et al. [857] made polyvinyl chloride surfaces with silanes and found that the most effective prevention of bacterial adhesion was obtained using 20% of dimethylsiloxane-co- (N,N-dimethyl-N-n-octylammoniopropylchloride)-methylsiloxane and (3 glycidoxypropyl) trimethoxysilane. Yoshino et al. [858] showed that coatings with silane and quaternary ammonium groups are stable for long periods of time. Gimeno et al. [178] demonstrated that antibiotic eluting systems are efficient in the prevention of biofilm formation on implants. A percutaneous porous wall hollow implant was used for an antibiotic eluting device *in vivo* model, and no biofilm was observed.

Electro-Assisted Methods

The advantages of electro-energy were used to demonstrate the prevention of bacterial adhesion. [108, 859, 860] The electro-assisted method utilizes the electro-repulsive interaction between bacteria and cathodic surfaces with a negative surface charge (Figure 24).

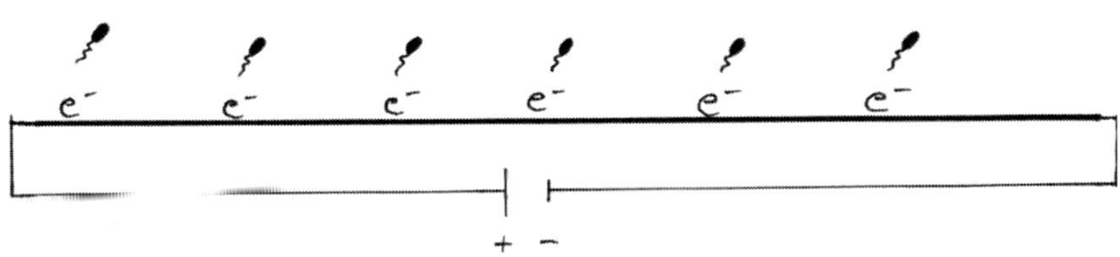

Figure 24. Electric current prevents biofilm formation.

Van Der Borden et al. [861] detached the initially adhered *S. aureus* biofilm from stainless steel with an application of less than 6.0 μA/cm² of cathodic or block current. In another study, Van Der Borden et al. [862] investigated the efficiency of 100 μA electric current and found electric current can prevent *S. epidermidis* biofilm formation in comparison to untreated pin tract.

The principle of using electrical current to prevent bacterial adhesion and biofilm growth was applied in boat technology to prevent marine biofouling. [860, 863, 864] There is a problem that when anodic current is applied inactivated cells remain on the surface and represent a possibility for the attachment of other cells. [865] The solution can be use of direct constant current or block current. Moreover, Wake et al. [864] showed that a block current is an effective antifouling strategy over a period of two years.

Cleaning and Maintaining Surface Hygiene

The most important measure that prevents bacterial proliferation and biofilm formation is a good cleaning process. [10] Use of high temperatures in cleaning can reduce the physical force required, such as water turbulence and scrubbing [202], but also biofilm removal. [866]

The process of cleaning removes more that 90% of bacteria, but afterwards those bacteria can become reattached. Cleaning can be used as prevention measure to restrain bacterial colonization and biofilm growth or as a strategy for control and elimination (see Chapter 9). Regular and efficient cleaning sustains a surface's hygiene. An example of a cleaning process can be seen in Figure 25. As a first step, surfaces should be rinsed with hot water to remove soil and dirt. In the next step, a chemical agent has to be applied to remove non-polar dirt e.g. oils and fats. However, chemical sanitizers will not remove all the soils; therefore, mechanical force has to be implemented to increase efficacy. After the second step, chemical agents have to be used for a second rinse with hot water to eliminate residues. In the final step, the surface should be dried as soon as possible to prevent bacterial growth (Figure 25). With continuous cleaning, surfaces are maintained in conditions that are not favorable for biofilm formation.

Afterwards, if cleaning did not remove all the soil, disinfection has to be applied to destroy cells' walls and to finally eliminate them. [867] The main strategy to prevent biofilm growth is to maintain clean surfaces before bacteria attach firmly to them. [642, 868] Some researchers have developed devices to monitor the surface colonization by bacteria. [869, 870] Pereira et al. [869] invented a mechatronic surface sensor able to detect biofilms in the early stages of development. Moreover, the sensor can detect the presence of cleaning products in a surface, identify it and evaluate the efficiency of cleaning. [871]

Other researchers attempted to provide a list of materials that prevent or suppress biofilm formation, but it is difficult find any material that does not provide any adhesion. [872-874] Frank and Chmielewski [875] studied the *B. stearothermophilus* and *Pseudomonas* spp. adhesion on stainless steel in the presence of milk and found that it was impossible totally prevent adhesion.

It is possible to prevent bacterial adhesion in unfavorable nutrient or temperature environments. [876] Several authors have reported that biofilms can be prevented via the incorporation of antibacterial products into surface materials [877, 878], by coating surfaces with antimicrobials [777, 879, 880] or by modifying the surfaces' physicochemical properties. [161, 881] Gottenbos et al. [777] provided implants with silicone rubber and quaternary ammonium coatings to prevent adhesion. Other authors demonstrated the inhibition of biofilm formation by coating surfaces with silver. [882, 883]

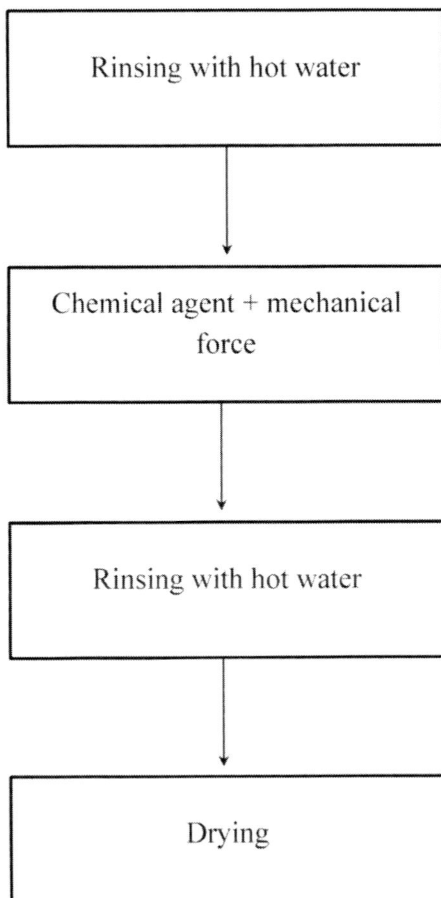

Figure 25. Process of cleaning for prevention of biofilm growth.

Uno et al. [884] studied biofilm formation on contact lenses under poor hygiene behavior and determined that contact lens care solutions can substantially decrease *S. epidermidis* biofilms. Hygiene cleaning practices were also tested by Wu et al. [885], who analyzed rinsing, rubbing, air-drying and tissue-wiping techniques. Mechanical rubbing and wiping were found to be the most effective techniques removing biofilm. Disinfecting the lenses in solution for six hours removed the majority of biofilms. Rinsing lens cases alone provided only minimal efficacy in reducing biofilm. Air-drying without any other additional cleaning methods was the least efficient at removing biofilms. An interesting study of Willcox et al. [886] showed that heating polypropylene contact lenses for three hours at 60 °C significantly reduced *S. aureus* and *P. aeruginosa* biofilms. Retuerto et al. [887] studied different disinfection strategies for contact lenses and found hydrogen peroxide to be efficient against biofilms. Hygiene behavior is the most important factor for the prevention of infection related to the contact lenses use. Antimicrobial lenses cases, efficient case cleaning, case replacement, and not using tap water to rinse cases are important measures for infection prevention. Patients should be informed how to clean and disinfect their lens cases, to avoid the use of tap water, to never top up their solutions and to replace them regularly. [888]

Adhesion can be also prevented by using surfactants as preconditioning films. [536] Different types of surfactants were studied for the prevention of *P. aeruginosa* biofilms. The results show that surfactants can prevent more than 90% of adhesion. Several different studies promote surfactants as a preventive strategy against biofilm formation. [889-891]

CONCLUSION

Sanitation programs must have a balanced combination of physical and chemical methods to assist in the cleaning process. The selection of suitable cleaning methods and cleaning agents are vital steps in the cleaning process. The aim of a sanitation program is to assure a hygienic environment. This program must be implemented for equipment and each area of the facility. It is essential that the program be designed to accomplish safety needs and remain economically feasible while meeting regulatory requirements. Good sanitation will be rewarded with improved work atmosphere, better productivity and a reduced chance of adverse incidents. All of these are reasons that it is important to establish and develop an effective sanitation program.

Chapter 9

Control of Biofilm Growth

"The disinfectant most effective to planktonic cells is not necessarily the most active against the biofilm"
(R. Van Houdt and C.W. Michiels)

Bacterial cells in biofilms differ from those in planktonic solutions because of several mechanisms that they have evolved in order to survive under various environmental conditions. Therefore, taking into consideration all microorganisms in various environments, control of biofilm growth is quite difficult, even if such control is possible against specific target species.

Nevertheless, effective chemical, physical and biological methods for controlling biofilm formation or microbial adhesion have been developed (Table 3). [82] Strategies to improve the hygiene status must not ignore the existence of bacterial sources, particularly on some equipment and floors, that are potential reservoirs and cause dissemination. This should be taken into consideration in order to design materials and coatings that are easy to clean and accessible for hygienic operation. [524]

Chemical Methods

Various chemical substances can react with many cell structures, e.g. cell walls, membranes, enzymes, genetic material and cytoplasm, causing their destruction or making them unable to reproduce. It is generally accepted that Gram-negative bacteria are better protected against chemical agents due to their thick cell walls than Gram-positive bacteria are. [501] In addition to the properties of bacteria, the properties of chemical agents play an important role in disinfection. Substances that are soluble in water and have around 600 Da molecular weight are effective because they can penetrate cell-wall pores. [10, 892] In everyday use, different types of chemical agents are used, which can be divided into different groups according to their mode of action: oxidizing agents; surface-active agents and iodophores. How efficient the disinfection will be against biofilms depends mainly on the concentration, pH, temperature, time and the presence of organic substances. [893, 894]

Table 3. Chemical, biological and physical methods for biofilms control

Methods	Agents
Chemical methods	Peracetic acid
	Ozone
	Sodium hypochlorite
	Hydrogen peroxide
	Chlorine
	Ethanol
	Metal ions
Biological methods	Enzymes
	Peptides
	Phages
	Natural antimicrobial substances
Physical methods	Mechanical cleaning
	Radiation
	Ultrasound
Combined methods	

Peracetic Acid ($C_2H_4O_3$)

Peracetic acid is an organic compound, a colorless liquid with a strong odor. Typically it is used in concentrations between 5% and 15%, especially for the disinfection of water. [5] Peracetic acid is a strong oxidizing agent that cannot be deactivated by catalase and peroxidased enzymes. It also has some environmental advantages since it can be applied without rinsing. However, due to its strong oxidative nature, peracetic acid cause corrosion. Frank et al. [895] reported that a concentration of 0.1% and exposure 10 min can reduce *L. monocytogenes* biofilm by 6 log. Other studies also show peracetic acid to be effective against biofilms. [896-898] Carmona et al. [899] examined porous silica loaded with peracetic acid against *S. aureus* biofilm and found it to be a highly efficient disinfection agent Lalueza et al. [900] found silver zeolite and peracetic acid to be an effective combination for the elimination of *S. aureus* biofilm; *P.aeruginosa* biofilm on a polystyrene surface was totally eliminated by 1.6% peracetic acid. [901] Park et al. [902] studied the inactivation of *S. typhimurium* and *L. monocitogenes* on stainless steel and polyvinyl chloride coupons with sodium hypochlorite and peracetic acid and obtained the best result for peracetic acid. However, other authors reported that peracetic acid is not significantly effective. [903-905] Mosteller and Bishop [906] showed the poor efficiency of peracetic acid against *P. fluorescens, L. monocytogenes* and *Y. enterocolitica* biofilms. Similar results were obtained by Rossoni et al. [904] who reported sodium hypochlorite to be more effective.

Ozone (O_3)

Ozone or trioxigen is a pale blue gas with a strong odor and oxidizing properties. [907] It is an antimicrobial agent that can be used against various kinds of bacteria, fungi and viruses. [908] Ozone causes lysis of the bacterial cells, which is much more efficient than transport through cell membranes in inactivating the bacteria. [909] There are several studies that indicate that ozone is an effective antimicrobial agent against biofilms. [4, 910, 911] Nevertheless, Tachikawa et al. [912] demonstrated that *P. fluorescens* and *P. aeruginosa* biofilms can become resistant against ozone. Ölmez and Temur [563] demonstrated that a concentration of 2 mg/L treatments at 10 °C for 2 min removed *E. coli* and *L. monocytogenes* from the surface of lettuce leaves. Di Ciccio et al. [913] reported that gaseous ozone treatment of *S. aureus* and *L. monocytogenes* resulted in a reduction of 0.82 and 1.39 log units for *S. aureus* and *L. monocytogenes*. However, there are some issues related to the ozone implementation regarding the properties of materials, temperature contact time and pH. [914] Ozonized water can be used to control biofilm grown in purified water circulation and storage. [915] Ozone was also shown to be a highly effective disinfection agent in the winemaking industry. [916] Ozone therapy is a new approach in dental hygiene and can be used for several diseases of soft tissue and for pits, fissures and caries. [916]

Sodium Hypochlorite (NaClO)

Sodium hypochlorite is a bleaching agent with strong oxidizing potential and strong odor. Some studies reported that sodium hypochlorite is effective against bacterial biofilms. [898, 917-919] It is also known to be effective at low pH levels. [920] Lomander et al. [921] showed that a 50 ppm concentration of sodium hypochlorite was effective against biofilms. Simon et al. [922] analyzed the efficiency of sodium hypochlorite in concentrations of 1 mg L^{-1} and reported a 78–84% reduction in adenosine triphosphate. In addition, a 6% sodium hypochlorite solution significantly reduced the biofilm in a bovine root canal model. [923] Zand et al. [924] studied the antibacterial efficiency of sodium hypochlorite on an *E. fecalis* biofilm and found that a 1% solution will eliminate more that 80% of the biofilm population. Similar results were reported by Norwood and Gilmour [925], showing that sodium hypochlorite was able to eliminate *L. monocytogenes, P. fragi* and *S. xylosus* biofilms. Biofilms seem to be more susceptible to sodium hypochlorite when they are under stress, e.g. starvation. [926] However, Capita et al. [927] studied *E. coli* after exposure to sodium hypochlorite and found adaptation of bacterial cells in the context of changes in morphology and ultrastructure, which allows biofilms to persist in the presence of sodium hypochlorite.

Hydrogen Peroxide (H_2O_2)

Hydrogen peroxide is a commonly used antimicrobial agent to control bacterial cells and biofilm formation. Hydrogen peroxide-based disinfectants are widely used in hospital environments to control bacterial colonization and contamination, and to reduce the risk of hospital-acquired infections. Some authors have reported hydrogen peroxide to be one of the most effective disinfectants. [898, 928] The advantages of hydrogen peroxide use are its low allergy properties and the fact that there is no negative impact due to product quality and affordability. [929, 930] Hydrogen peroxide has been found to be effective in low concentrations, (e.g. 0.05%), and a 0.2% solution can also eliminate mature biofilms. [931] Robbins [932] reported that a 5% solution in 15 min exposure eradicates all biofilms Stojicic et al. [933] demonstrated that biofilms are susceptible to hydrogen peroxide, but after three weeks resistance appeared with only 30% efficiency. Perumal et al. [934] compared the efficiency of hydrogen peroxide-based disinfectants on planktonic cells and biofilms of *K. pneumonie* and *P. aeruginosa* and found that biofilms are more than two hundred times more resistant to hydrogen peroxide. Chaieb et al. [935] noticed that 0.1% hydrogen peroxide is not efficient in hospital environments, and emphasized that this can be one reason for hospital-acquired infections.

Chlorine (Cl_2)

Chlorine is commonly used in the disinfection of water and water systems, for which four orders higher concentrations are required for the elimination of biofilms in comparison to planktonic cells. It has broader pH range activity compared to sodium hypochlorite. [936] The disadvantages of chlorine application are corrosion due to its high oxidizing properties and toxic components in the presence of organic compounds. [937-939] Norwood and Gilmur [925] reported that 1000 ppm and more is required for biofilm elimination, while for planktonic cells only 10 ppm is required. One of the reasons for biofilm resistance is (beside the low diffusion) the chlorine reaction with organic substances at the surface of the biofilm, which decreases the efficiency of penetration. [938-941] Nam et al. [942] studied the impact of gaseous chlorine for inactivating *B. cereus* spores and determined a lethal dose at 115 ppm within six hours. Schlisselberg and Yaron [943] studied the sensitivity of *S. typhimurium* to chlorine on different stainless steel finishes and found it to be was the most effective on electro-polished surfaces. Tinetta et al. [944] studied the disinfection of commercial meat slicers and an industrial hot dog peeler when contaminated with *L. monocytogenes* biofilm and found 2 mg/L to be efficient for inactivation. Gao and Liu reported that some strains of *L. monocytogenes* can become chlorine resistant after repeated exposure to chlorine at sub-lethal levels. [945] O'Leary [946] found that strains firmly attached to stainless steel and plastic surfaces survive high chlorine concentrations.

Ethanol (C_2H_5OH)

Ethanol is the most common alcohol used in disinfection. [947] The advantage of ethanol disinfection is its non-corrosive properties and affordability. However, it can be a fire hazard and has limited residual activity due to evaporation, which results in brief contact times. Ethanol is most effective when diluted with water to facilitate diffusion through the cell membrane. [948, 949] A 96% alcohol solution typically denatures only external membrane proteins. Møretrø et al. [950] tested nine different disinfectants against *Salmonella* spp. and concluded that 70% ethanol was most effective. In contrast, Milisavljevic et al. [951] incubated *S. epidermidis* in ethanol and found that induction or increases in biofilm production. Chauhan et al. [952] reported that ethanol can inhibit *C. albicans* biofilm formation. Walker et al. [953] studied different disinfectants on mixed species biofilm and found that ethanol-based disinfectants did not remove biofilms. Korem et al. [954] reported *S. aureus* exposure to the ethanol induced oxidative stress and increased pathogenic traits.

Metal Ions

Several studies reported the significant use of silver ions against bacteria, fungi, protozoa and viruses. [955-957] Silver has several implications in medical devices, such as impregnated dressings and catheters. [882, 958-961]

It has also been proposed that the impregnation of silver into a coating can be more effective than direct coating alone, since surface silver can be readily deactivated by protein anions. [955, 958] Silver zeolite was found to be effective against bacterial growth on tooth implants. [962] Baker et al. [963] reported that nano-particles are efficient against *E.coli* biofilms. Kong and Jang [964] studied the antibacterial properties of silver nano-particles and sulfadiazine and silver nitrate against *E.coli* and *S.aureus* and found the best results for nano-particles. Harrison et al. [965] reported that several ions, such as CrO_4^{2-}, Co^{2+}, Cu^{2+}, Ag^{+}, Zn^{2+}, Cd^{2+}, Hg^{2+}, Pb^{2+}, AsO^{2-} and SeO_3^{2-}, have an impact on *C. albicans* biofilms. However, the use of medical devices containing silver must be used with caution since silver can be toxic for humans. [966] Tabrez Khan et al. studied the impact of copper and zinc oxides to biofilm formation in oral cavities and found that a concentration of 50 μg/L significantly inhibits all oral bacteria. [967] Ancion et al. [968] studied the impact of different heavy metals on bacteria and found a 73% reduction for copper, 48% for zinc and 46% for lead. In another study, Hu et al. [969] examined the effects of Cu^{2+} ions on EPS formation and found that concentrations above 5 mg/L cause biofilm instability. Metal oxides, including CuO and Fe_2O_3, were tested for antimicrobial activity against *S. aureus* and *E. coli*; CuO proved to be more toxic than Fe_2O_3. [970] Titania was found to be very efficient against biofilms. [971] Chorianopoulos et al. [972] used titanium for disinfection food-processing plant and found 16% w/v to be effective against *L. monocytogenes.* Similar potentials of titania were found by Pozo et al. [973] for *P. fluorescens* on polystyrene surfaces. Dunlop et al. [974] reported that photocatalytic disinfection with titania is effective against *S. epidermidis*.

Adding metal ions to the polymer surface promotes the formation of active oxygen species from peroxides and persulphates, thereby targeting the cells nearest to the surface. This local antibacterial treatment affects the cells on the surface and prevents adhesion. [975]

BIOLOGICAL METHODS

In addition to classic methods of biofilm control, nowadays more ecologically friendly approaches are becoming important. Biological methods have been recognized as promising candidates for the next generation of biofilm control. Advantages of this approach include a low propensity for developing bacterial resistance, broad spectrum activity, high efficacy and target specificity. [10]

Enzymes

Enzymes are proteins with catalytic activity and are widely used in the food industry. [10] The efficiency of enzymes for biofilm control is bacterial specific; therefore, the correct enzymes for each bacteria have to be used. Combinations of protease and amylase seem to be effective against various biofilms. [976]

A study by Molobela et al. [232] showed highly effective use of protease enzymes against *P. fluorescens* biofilms, while amylase enzymes were less effective. α-amylases are commonly used because they are thermo-resistant, but their disadvantage is that they have a short time of efficacy as they are calcium metallo-enzymes. [977] Orgaz et al. [978] reported that enzymes, such as pectin esterase, pectin lyase and cellulose, are not effective against mature *P. fluorescens* biofilms. Dispersin B is another enzyme used for the eradication of *S. aureus, S. epidermidis, E. coli,* and *Y. pestis* biofilms. [273, 979-981] An *in vivo* study showed that Dispersin B has the potential to prevent the formation of *S. aureus* biofilm in catheters. [982] Alginate lyase was shown to be effective against *P. aeruginosa* biofilms *in vitro* [983, 984], and *in vivo* in a rabbit model [985]. Enzymatic cleaning agents are also used in the dairy industry against several bacterial strains, including *L. bulgaricus, L. lactis* and *S.thermophilus.* [986] However, the authors also reported that efficacy was decreased by the presence of milk and other organic matter.

The specific action of enzymes and the specific properties of bacterial cells make it difficult to create an enzyme-based cleaning agent for all bacteria. Formulations containing several different enzymes seem to be fundamental for a successful biofilm control strategy. Principally, proteases and polysaccharide hydrolyzing enzymes may be useful. [876]

Peptides

In recent years, natural antimicrobial peptides have attracted considerable interest as a new antimicrobial strategy. The advantages of antimicrobial peptides are their broad activity spectrum, selectivity towards microbial membranes, their rapid mechanism and, most importantly, the low frequency in selecting resistant strains. [987]

Guerra et al. [988] showed that nisin, an antimicrobial peptide also used as food preservatives, adsorbed to stainless steel, rubber and polyethylene terephthalate surfaces and, upon doing so, retained its antibacterial activity and inhibited the growth of *E. hirae*. Tong et al. [989] found nisin to be efficient against the cariogenic bacteria *S. mutans*. Moreover, nisin-coated polyethylene terephthalate bottles significantly reduced the total aerobic plate counts in skim milk compared to uncoated bottles, although it was not clear whether the effect was because of adsorbed nisin or nisin released into the milk. Nevertheless, this polyethylene terephthalate-based bioactive packaging extended the shelf-life of the milk and consequently could be a promising technique for extending the shelf-life of various packaged foods. [988] Moreover, Malcolm et al. [990] found that the secretion of antimicrobial proteins and salivary antibodies can impact biofilms in the oral cavity.

Phages

Bacteriophages are viruses that can infect bacteria and prevent their growth. [991] The first application of phages was in the early 20th century for the treatment of bacterial infections, but some researchers have shown that phages are also suitable for biofilm control. [992, 993] Phages have advantages in that they are non-corrosive, unlike chemical substances. [994, 995] Some authors have reported that the production of phages is strongly influenced by bacterial growth and environmental conditions. Corbin et al. [996] showed that a T4 phage is effective against *E.coli* biofilm even at the nutrient limit. Sillankorva et al. [997] studied the efficiency of phage philBB-PF7A on *P. fluorescens* biofilms in static and dynamic conditions and found better (i.e. higher) biofilm destruction under static conditions. Cerca et al. [998] demonstrated that the action of staphylococcus bacteriophage K depends upon the bacteria growth phase. In contrast, Tait et al. [999] reported that phages and bacteria can co-exist in biofilms; therefore, a mixture of phages, depolymerases and disinfectant was suggested for better biofilm control. With the use of biotechnology principles, phages can be engineered to express degrading enzymes, which has potential for biofilm control. The engineered phages were reported to be able to totally eradicate biofilms. [1000]

Natural Antimicrobial Substances

Essential oils are derived from plants and are recognized as safe in comparison to classical antimicrobial agents. [1001] Several essential oils have antimicrobial properties, including bergamot (*Citrus bergamia*), cinnamon (*Cinnamomum verum*), ginger (*Zingiber officinale*), lavender (*Lavandula stoechas*), lemongrass (*Cymbopogon citratus*), oregano (*Origanum vulgare*), pine (*Pinus sylvestris*), and rosemary (*Rosemarinus officinalis*). [1002-1004]

The terpenoids and phenolic compounds can penetrate the cell wall, causing protein denaturation. [1003, 1005-1007] The essential oils also decrease the pH levels, which results in membrane disruption and the loss of control of DNA transcription, protein synthesis and enzyme activity. [1008] Several studies have shown that essential oils are effective against bacterial biofilms, such as *E. coli, L. monocytogenes, S. aureus, S. enteritidis* and *S. Typhimurium.* [1003, 1004, 1008-1010]

Chitosan is a polysaccharide that can be derived from the chitin of crustacean shells, such as crab shells, fungi cell walls and other biological materials. Chitosan was found to be effective against *E. coli, S. aureu, L. monocytogenes, B. cereus, S. Typhimurium, Penicillium, Aspergillus.* [1011-1014] Chitosan impacts cell membranes due to their positively charged molecules, which interact with negatively charged bacteria, causing cell disruption. [1011, 1013, 1015]

Another group of natural antimicrobial substances are organic acids and their salts, including potassium sorbate, sorbic acid, p-aminobenzoic, malic acid, lactic acid, propionic acid and others. Several researcher have shown that these substances are effective against biofilms, such as *E. coli, L. monocytogenes, S. Typhimurium, S. aureus, B. cereus, A. niger* and *A. flavus.* [1005, 1013, 1015, 1016] The main mechanism of antimicrobial activity is the disruption of cell walls transport due to low pH. [1015-1017]

Nisin is an antimicrobial peptide produced by *L. lactis* and is often used in the food industry. [1007, 1013] However, it seems that nisin have no effect on Gram negative bacteria. [1013, 1018] Its mode of action includes the inhibition of cell wall synthesis and the formation of pores in the cytoplasmic membrane. [1018] Some researchers have proven nisin to be an effective control *against L. monocytogenes, S. aureus and B. cereus.* [1007, 1013]

Physical Methods

Physical methods are one of the oldest techniques of biofilm removal. Biofilms can be eliminated by scrubbing, scraping, high-pressure use, ultrasound or radiation. The greatest advantages of physical methods are that bacteria cannot develop resistance against them and that they are relatively cheap and easy to use.

Cleaning Processes

The purpose of cleaning is the removal of product residues, dirt and bacteria. Mechanical cleaning is probably the oldest and most successful method; with force, organic matter is removed from surfaces. When biofilms are fully mature, only extensive cleaning processes can eradicate them. In addition to preventive cleaning (see Figure 25), cleaning for the control of biofilm also includes the disinfection process (Figure 26).

In addition to mechanical action, other physical techniques can also increase the efficiency of biofilm control. Boulange-Petermann et al. reported that ultrasonic treatment is efficient for cleaning in the food industry. [1019] Wu et al. [1020] proposed the use of small bubbles for biofilm detaching. Similar principles are used for cleaning water pipes. [1021, 1022]

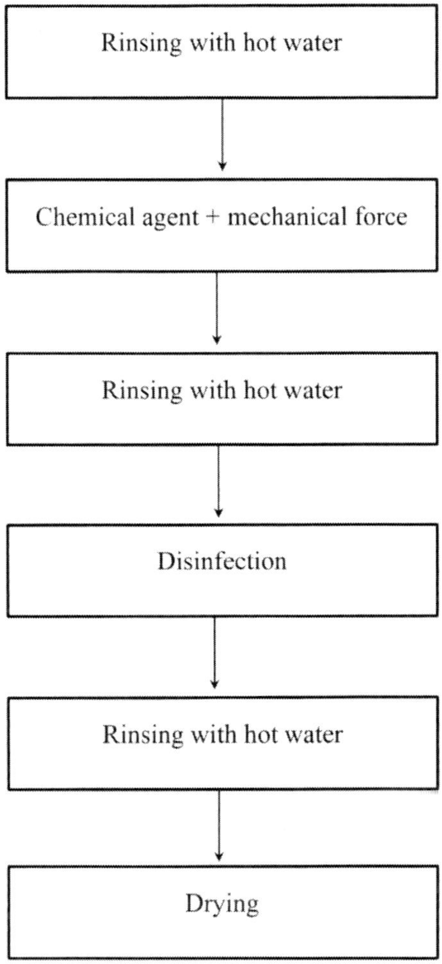

Figure 26. Process of cleaning for the control of biofilm growth.

The problem of with this manner or cleaning is that biofilms are not equally distributed on the material, and the cleaning has to be focused in certain sites or fittings. [1023] Therefore, it is important to have information of the location of the biofilm. In order to pinpoint the location of the biofilm, a systematic sampling approach is required. Access to surfaces is a great advantage and should be taken into the account at stage of planning.

For hygiene, it is important to break down the biofilm structure and remove the entire biofilm [24], since incomplete removal causes the reattachment of bacteria to the surface and the formation of a new biofilm.

[70] Furthermore, disinfectants are less effective when dirt particles are present on surfaces. [201, 1024] Common cleaners, such as alkali-based and acid-based cleaning, are only adequate in removing the EPS matrix when the correct formulations, concentrations, time, temperature and flow are applied. It must be realized that suboptimal process parameters will drastically affect the efficiency of cleaning. [59, 1025, 1026]

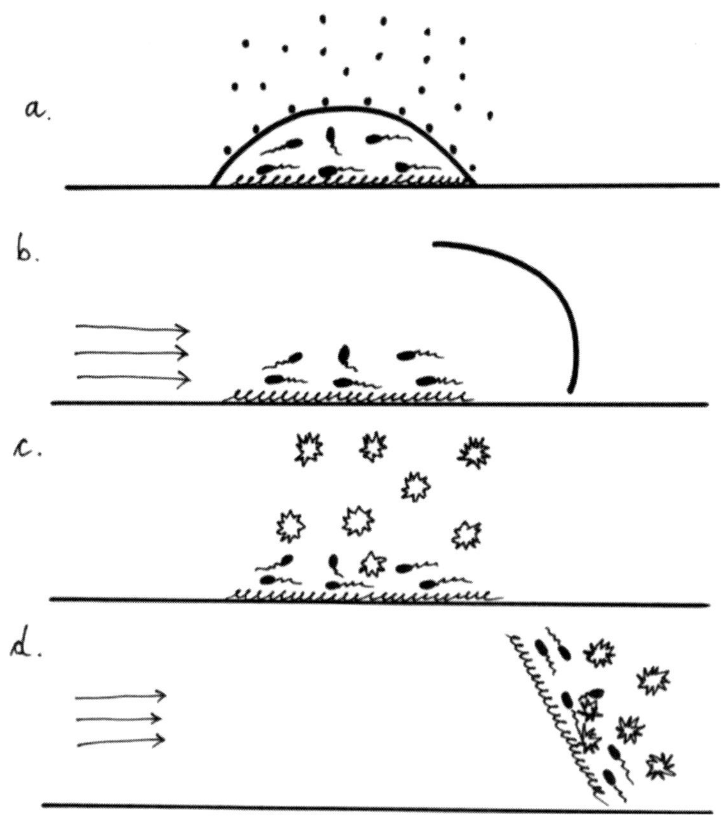

Figure 27. Four steps of biofilm removal.

The cleaning must be done in such a way that the procedure dissolves the EPS matrix and assures access to bacteria cells [868] (Figure 27a). In the next step, rinsing with water will remove the dissolved EPS matrix and increase the efficiency of the third step (Figure 27b). The cleaning process can remove most bacteria if they are not present in a mature biofilm; therefore, disinfection must be implemented after the cleaning; otherwise, the organic substances will interfere with the disinfectant (Figure 27c). [566, 867, 1027] After the disinfection, dead bacterial cells, along with the rest of the EPS matrix and conditioning film, should be removed by rinsing with water (Figure 27d).

A good disinfectant must be effective, safe, cheap, and easy to use, ecologically friendly and not leave toxic residues that could affect the sensory values of the final products. [1028, 1029]

Sadoudi et al. [1030] reported that pulsed laser beams could be used as an alternative cleaning method for the reduction of bacterial populations. However, a disadvantage of this method is that it cause aerosols, and some extra measures should be taken to prevent spreading microorganisms. This is one of the reasons the use of high-pressure sprays has been replaced by foam cleaning. A study by Gibson et al. [70] reported that use of high-pressure spray at distances greater than 250 mm from the surface is not recommended, because this will cause the generation of aerosols and will enhance the dispersion of the bacteria. The author also reported that pressure above 17 bar does not remove biofilm formations from surfaces.

Cleaning out of Place (COP)

Parts of the equipment that are difficult to clean should be cleaned with COP. In this case, equipment must be taken apart or moved from the production area to a cleaning area. [1031] One of the advantages of a COP system is that it can efficiently clean different parts of equipment and utensils. The system is more cost effective and hygienic than washing by hand. [1032] The units for the COP system have sanitary tank with a recirculating pump and distribution headers for the efficient distribution of cleaning agents.

Cleaning in Place (CIP)

CIP is a method of cleaning the interior surfaces of pipes, vessels, process equipment, filters and associated fittings, without disassembly or manual operations. [976] The CIP procedure has the following steps: a pre-rinse with cold water to remove gross residues; the circulation of detergent; an intermediate cold water rinse to flush out detergent; the circulation of disinfectant; a final cold water rinse to flush out detergent (Figure 28). [1033]

However, some authors reported bacterial residues and biofilm formation after CIP. [44, 1034, 1035] Dufour et al. [1036] investigated different CIP regimes for dairy biofilm control with the use of chlorine, nisin and lauricidin, and found that the measures are not totally effective against biofilms. Similarly, Bremer et al. [1034] also reported the low efficiency of a standard CIP regime (water rinse, 1% sodium hydroxide at 65 °C for 10 min, 1% nitric acid for 10 min, water rinse). Biofilm evaluation and control should be part of the HACCP (Hazard Analysis Critical Control Point) development plan in order to control those biofilms prevalent in the processing areas. [397, 1037, 1038]

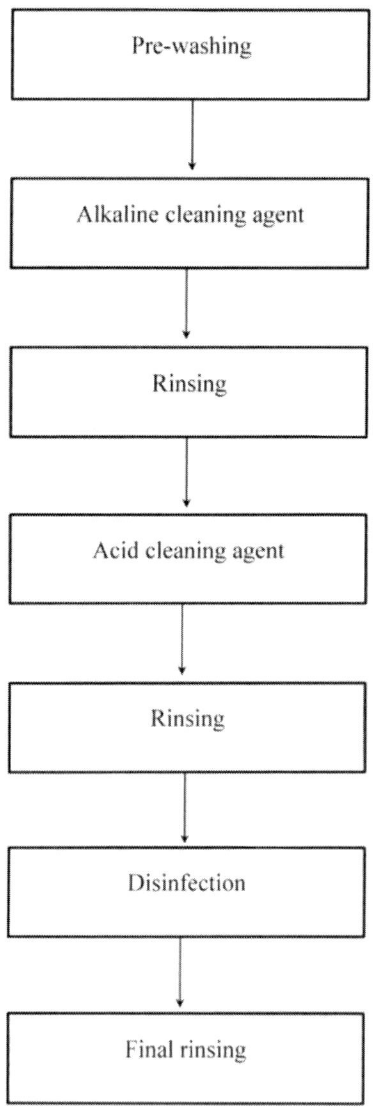

Figure 28. Process of CIP.

There are many factors that can influence CIP efficacy, including the type of biofilm layer, the chemical composition and concentration of agent, time, temperature, hydrodynamic, and the cleaning surface characteristics [1019, 1034, 1039].

Radiation

An increase in temperature due to radiation accelerates biochemical reactions. Proteins exposed to temperature leads to denaturation and an interruption of the process.

Niemira and Solomon [627] proved that ionizing radiation is effective against *Salmonella* spp. biofilms. In another study, Niemira [1040] examined gamma radiation as a potential strategy of biofilm control. The results show that *L. monocytogenes* is sensitive to ionizing radiation, while for *E. coli* it was found that mature biofilms are sensitive as young ones. Khaw et al. [1041] emphasized that mucositis and peridonitis can be treated by radiation. Similar conclusions were published by Feuerstein [1042] when studying light therapy as complementary antibacterial treatment of oral biofilm. Bernbom et al. [1043] reported on bactericidal effects on *L. monocytogenes* in food-processing plants with UV-C light. The authors also emphasized that in the presence of sodium chloride, organic food materials decrease the effectiveness of radiation. Some progress in the field of ionizing radiation shows that atmospheric plasma inactivation makes use of reactive oxygen species and radicals that can effectively remove biofilms. Vleugels et al. [1044] reported that atmospheric plasma is effective against mature biofilms and planktonic cells. Oulahal-Lagsir et al. [1045] tested a synergistic treatment of ultrasound and enzymes and reported the efficient removal of *E. coli* biofilms. Pezzoni et al. [1046] studied the impact of UV-A radiation on *P. aeruginosa* and found that catalase enzymes can protect bacteria cells against irradiation.

Ultrasound

Ultrasound can be used to remove bacterial population from surfaces or to increase the efficacy of disinfectants. [1047-1050] In recent years, the use of ultrasound for the elimination of biofilms has received increasing attention. The process includes the formation of microbubbles, where their growth and collapse occur in a short time with the release of intense and focused energy.

[1051] Oulahal-Lagsir et al. [1049] reported that ultrasound is an efficient method of removing biofilm from stainless steel and polypropylene surfaces. Nithila et al. [1052] reported that the application of 24 kHz and 400 W of ultrasound can efficiently eliminate *Pseudomonas* spp. biofilm. Similarly, Iqbal et al. [1053] studied the impact of high intensity ultrasound on *E. fecalis* and found that efficacy is dependent on time exposure. Some authors also reported that low frequencies are more efficient than high ones. [1054-1056] It is also known that ultrasound can increase the effectiveness of antibiotics against biofilm cells. [1057] One reason for this can be in the higher diffusion of oxygen into the biofilm matrix, causing the cells to become active and, therefore, more affected by antimicrobial agents. [1058, 1059] Baumann et al. [1048] also reported a significant effect on biofilm removal on stainless steel surfaces via the use of ozone and sonication. Similarly, Thurnheer et al. [1060] studied the ultrasonic removal of mixed species biofilm populations and observed the detaching of bacteria from the surface. Agarwal et al. [1061] proposed a chemical-free method with low intensity ultrasonication for removing biofilms from membrane with an efficacy of 70%.

Combined Technologies

A hurdle technology is a combination of two or more different control techniques to eliminate the biofilms. The aim of combined technologies is to use the advantages two or more sanitizers to increase efficiency via a synergistic effect. [646, 1062] With hurdle technologies, the first step removes EPS or increase the penetration of the cleaning agent. This pre-treatment increases the efficiency of the cleaning process. In the next step, the biofilm is deactivated by a cleaning agent or disinfectant so that it can be totally eliminated in the final step (Figure 29)

Figure 29. Steps in hurdle technology for biofilm elimination.

Ha and Ha [1063] demonstrated that a combination of sodium hypoclorite and ultraviolet irradiation had a better elimination of bacteria cells than a single treatment. DeQueiroz and Day [1064] reported on the effectiveness of a combination of sodium hypochlorite and hydrogen peroxide in eliminating *P. aeruginosa* biofilms. Similar results were determined by Vankerckhoven et al. [1065], showing a synergistic effect of a combined treatment of biofilms with hydrogen peroxide and ultraviolet irradiation to be ten times more effective than the single use of hydrogen peroxide. Schurman et al. [1066] analyzed the effectiveness of hydrogen peroxide with increasing temperature and demonstrated a synergistic impact. Additionally, the synergistic effect of ozone and ultrasound was also shown to be efficient for biofilm cell reduction. [1067] Moreover, Verma et al. [1068] found that a hurdle technology of antibiotics and bacteriophages to be more efficient that a single technology against *K. pneumonia*. Nguyen et al. [531] studied the impact of temperature and pH level on *S. typhimurium* formation on a stainless steel surface and reported that a hurdle technology of low temperatures and reduced pH may delay biofilm formation on food contact surfaces contaminated with *S. typhimurium*. Haaken et al. [1069] demonstrated that the disinfection of wastewater was more effective in a combination of ultra violet irradiation and electrolysis than by a single technology. Torlak and Sert [1070] demonstrated *L. monocytogenes* biofilm removal via a combined technology of ultrasound and bezalkonium chloride. [1055] The antimicrobial efficacy of ozone can be enhanced by combining it with hydrogen peroxide or ultraviolet radiation exposure. For many chemical approaches, mechanical action is also needed to dislodge microorganisms from the surface and expose them to antibacterial agents.

In many studies, enzymes and detergents have also been used as synergists to improve biofilm elimination. [894, 1025, 1071] Furthermore, Parkar et al. [1025] showed that the combination of enzymes and surfactants increased wettability and, therefore, enhanced cleaning efficiency.

FACTORS AFFECTING ANTIBACTERIAL ACTION

Proper cleaning is one of the most important measures to control the life of a biofilm on a surface. The main environmental factors that influence the activity of a biocide are chemical formulation, temperature, pH, water hardness, and the presence of additives. [67, 1072, 1073]

The efficiency of disinfectants depends on the temperature; in general, it increases with rising temperatures (Figure 30). When disinfection occurs at low temperatures, the use of higher concentrations of disinfectants or longer contact time may be needed to increase effectiveness. Simões et al. [10] reported if substances, such as carbohydrates, fat, proteins, blood residues or dirt or present, disinfection is less effective. Such contaminants may have a significant impact on cleaning and disinfection [32] (Figure 30). The disinfection process is dependent on the concentration, which is normally given by the manufacturer based on laboratory tests analyzing the efficacy of suspension against microorganisms (Figure 30). Disinfectants, such as alcohols, usually lose efficacy with dilution, while chlorhexidine, glutaraldehyde, ortho-pthalaldehyde retain much of their activity with dilution. [1073, 1074]

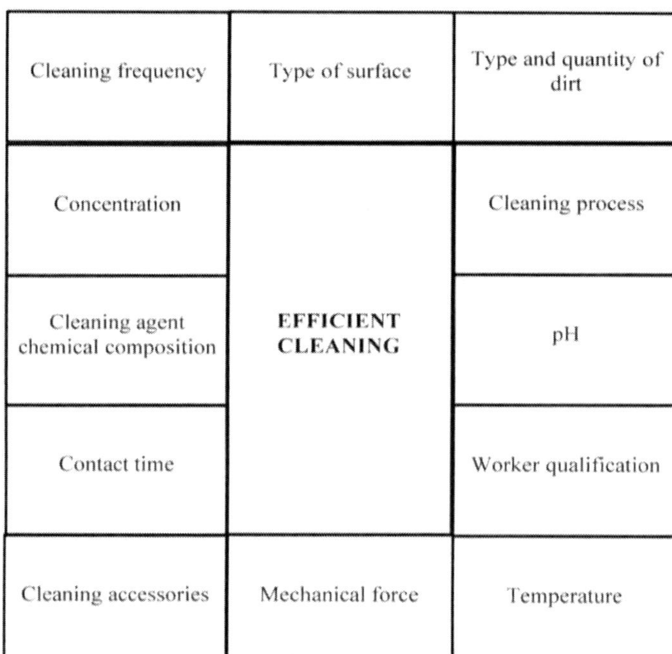

Figure 30. Factors determining efficient cleaning.

In addition to all these factors, the type of surface and type of bacteria should also be taken in to account when planning disinfection. A standard disinfectant should kill 5 log of bacteria in 30 s. [1075]

Although increasing the concentration will result in greater effectiveness of disinfection, in some cases (e.g. disinfection of foodstuffs) this can result in low product quality [562]

Bacterial cells on foodstuffs can be protected by stomata, trichomes, calyx, and stems, [563, 564, 1076] making disinfectants unable to access cells [553]. In these cases, the standard cleaning procedures cannot efficiently remove the cells. [563, 564, 1076] Brandl and Mandrell [1077] reported that *S. thompson* was present on cilantro plants and that disinfection was not efficient due to the disrupted cuticles of the damaged regions. Similarly, other authors have found that hydrophobic surfactants are unable to eliminate bacteria from the cutting sites of lettuce. [1078] It is well known that high temperatures and humidity cause vulnerable conditions for food-borne pathogens, as well as forming biofilms and reducing the effectiveness of disinfectants. [1079, 1080]

Furthermore, the period from when the first contamination occurred is important. The longer the time is, the more firmly are do the cells adhere. Sapers [559] highlighted the fact that after 24 h, *E. coli* on apples are so firmly attached that washing with 1% hydrogen peroxide was not efficient. Other researchers also proved that biofilms are strongly attached to the surface and are very difficult to remove with cleaning agents. [3,12,1081] or [1082] Joseph et al. [941] showed that 10 ppm of chlorine kills 6 log of *Salmonella* spp. planktonic cells, but only 1 log of biofilm. A similar result was reported for *S. aureus* biofilm, where concentrations of benzalkonium chloride and sodium hypochlorite were up to 600 times that needed for planktonic cells. [1083] Another study by Robbins et al. [932] showed that 0.25 ppm ozone decreased the number of *L. monocytogenes* planktonic cells by 8.3 log, while biofilms cells were decreased only by 1.5 log. With the maturation of a biofilm, the resistance to chemical agents is increased; therefore, increasing levels of sanitizers are used to achieve the same results. [1084-1086]

Most laboratory studies on biofilms are based on a single bacterial culture, but such cases do not appear in nature. [389, 1087] In nature, biofilms consist of multiple species. Biofilms of multi-species are more resistant to cleaning agents and disinfectants due to their complex architectures and inter-species and intra-species interactions. [1088-1090] Behnke et al. [1081] demonstrated that a mixed culture of *B. cepacia* and *P. aeruginosa* can be eliminated by chlorine of 80 ppm, while only 30 ppm is sufficient to kill a single species. Some of the research on multispecies biofilms found that such biofilms have individual capabilities and synergistic interactions, physiological heterogeneity and antimicrobial resistance. [610, 1091, 1092] Chorianopoulos et al. [1093] found that multispecies biofilms are more susceptible to cleaning agents and disinfectants than those of a single species. In multispecies biofilms, bacteria are in close contact, communicate via inter-species or intra-species QS, resist gene transfer within species and can, therefore, combat sanitizers as a whole. [209, 1088]

One of the most important recommendation is that equipment and materials should be hygienically designed so that bacteria cannot attach to the sites that protect them from cleaning. The European Hygienic Engineering & Design Group documents should be consulted in this respect. [69] Product contact materials must be inert both to the product and to detergents and disinfectants. They must be resistant to the corrosion, non-toxic and mechanically stable. The materials have to be smoothly finished and easy cleanable. Equipment with excellent hygienic characteristics but without the ability to perform its duties is of no use. Direct metal-to-metal joints other that welding should be avoided since such joints will harbor dirt and microorganisms. In addition, gaps due to misalignment in equipment and pipes must be avoided. Seals and gaskets must be designed in such a way that dirt and bacteria cannot accumulate. Corners should have radii larger than 6 mm. Sharp corners should be avoided. Product contact surfaces should have surface finishes of 0.8 μm Ra. Pipework must be self-draining and easily cleanable. Horizontal surfaces should be avoided; instead surfaces should slope to the one side. External surfaces should have drainage away from the main production area. Maintaining low air temperature and humidity should be provided to prevent bacterial growth. Cold spots where water condenses should be eliminated. Surfaces such as floors and walls can be indirect sources of bacterial contamination that can be consequently transferred to the product by vectors such as the air, personnel and cleaning systems. [893] The hygiene of the surfaces, therefore, affects the quality and safety of the food product.

CONCLUSION

Regular cleaning, disinfection and maintenance are common strategies to control pathogenic bacteria. However, such procedures are not very effective on biofilms and can induce the selection of resistant phenotypes; therefore, new emerging control strategies have to be implemented. Biological, chemical and physical methods, each with its own properties, can be used to overcome biofilm antibacterial resistance.

Chapter 10

FUTURE TRENDS AND CHALLENGES

"Vision without action is a daydream. Action without vision is a nightmare"
(Japanese proverb)

The microorganisms in biofilms live in self-produced extracellular polymeric substances that form their immediate environment. EPS are mainly polysaccharides, proteins, nucleic acids and lipids; they provide the mechanical stability of biofilms, mediate their adhesion to surfaces and form three-dimensional polymer networks that interconnect and transiently immobilize biofilm cells. EPS are the main reason that bacteria develop resistance to antimicrobial agents; therefore, future research will be focused on the control of EPS.

RESISTANCE OF BIOFILMS

The use of antimicrobial agents is a convenient method of controlling biofilm growth. Nevertheless, bacterial populations in biofilms are generally resistant to most antimicrobial agents that are effective against planktonic cells. [11, 1028, 1029] Kim et al. [930] reported that *L. monocytogenes* biofilms are the most resistant to sanitizers and disinfectants. Among antimicrobial agents, bacteria often produce resistance against trisodium phosphate, chlorine, ozone, hydrogen peroxide and peracetic acid. [618,932,1094] Which disinfectant is the most effective deepens on several factors, including the type of attachment, temperature, exposure time, concentration and pH. [1028, 1095]

Bacterial resistance is the result of different mechanisms. Slow or incomplete penetration of the antimicrobial agent into the film is most frequent mechanism of bacterial resistance (Figure 31). [32, 1096] The physiology of biofilm cells can be changed in such a way that cells express an adaptive stress response. One way to produce resistance is also by differentiation of small subpopulations into persisting cells of cells. [375]

Biofilm cells that are located deep within the biofilm are exposed to changed nutrient gradients. [1097] This causes a dormant state that increases resistance to biocides (Figure 31). [1098, 1099] Several authors have reported that mature biofilms are more resistant to antimicrobial agents than younger biofilms are. [1094, 1100] One example are *S. enteritidis* isolates that survived exposure to acidic conditions and the presence of hydrogen peroxide, which also showed enhanced resistance towards heat. [1086, 1101]

An explanation of resistance can also include the idea that biofilm cells can sense the biocide and respond actively by providing protective stress responses more effectively than planktonic cells do (Figure 31). [1102] Repeated exposure of *P. aeruginosa* biofilms to monochloramine resulted in a loss of effectiveness. [1103-1105] Another principle of increased resistance is the horizontal gene transfer of resistant plasmids in the extracellular DNA. [1106] This is also a metabolic explanation of resistance. [1107] Lewis et al. [1108] reported on metabolically dormant bacterial cells that exist in biofilms and allow for survival after antibacterial agents eliminate the majority of biofilm cells.

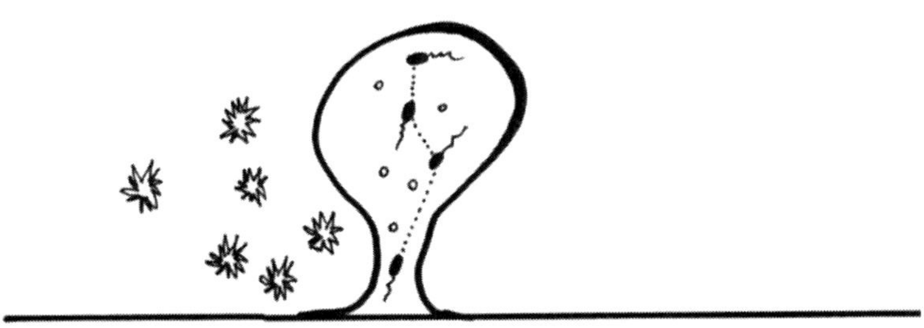

Figure 31. Mechanisms of biofilm resistance.

Since planktonic bacteria are more susceptible to antibiotics, the delivery antibiotics to prevent biofilm formation has been examined (see Chapter 8). Local delivery solutions have some advantages in relation to toxicity, the wasting of expensive antibiotics, and ensuring minimum inhibitory concentration.

In particular, several studies have found that bacteria in a biofilm can naturally produce agents that cause dispersal. [1109, 1110] Such chemical substances are d-amino acids, cis-2-decenoic acid, nitric oxide, peptides and enzymes. [1111-1114] Naturally occurring dispersal agents represent a new study niche for exploring new strategies for combating biofilms.

Bacterial Interference

Bacterial interference is a strategy that bacteria can use to inhibit one another when competing for resources in the same environment. This phenomenon may be used to prevent infections on surfaces where normal flora is present, e.g. dental implants, chatters, contact lenses. Trautner et al. [1115] demonstrated this principle by colonizing a catheter surface by non-pathogenic *E. coli*, thereby preventing the adhesion of pathogenic strains. Similar research on non-pathogenic strains preventing the colonization of pathogenic ones was also reported by Ishijma et al. [1116] Catuogno and Jones [1117] reported that the colonizing probiotic *S. salivarius* represented a potential mitigation strategy to prevent pathogenic biofilms.

Multi-Species Biofilms

Available research has studied the efficacy of disinfectants on monoculture biofilms, but such types of biofilms are rarely present in nature. [389, 1087, 1118] Biofilms in nature consist of multiple bacterial species causing less susceptibility to antimicrobial agents due to their complex architectures and interactions. [1088, 1090] Several studies have reported that bacterial strains in such biofilms work synergistically to overcome unsuitable environmental conditions. [610, 1091, 1119, 1120]

Behnke et al. [1081] reported that a mixed culture of *B. cepacia* and *P. aeruginosa* can be inactivated with 80 ppm chlorine, while only 30 ppm is enough to eliminate a single species. Similarly, Chorianopoulos et al. [1093] examined mixed culture populations of biofilms, tested their sensitivity to surfactants and found greater resistance for multi-species biofilms. In another study, Shu et al. [1121] analyzed oral species of wild and mutant strains of biofilms and proven multispecies interactions within microenvironments. In mixed species biofilms, bacteria interact even more intensely than in single species biofilm. They communicate by producing QS, resist gene transfer within species, and therefore resist sanitizers. [209, 1088] Lee et al. [1122] studied several bacterial strains (e.g. *P. aeruginosa*, *P. protegens*, and *K. pneumonia*) and found that single-species biofilms were resistant to one antibiotic, while multispecies biofilms were resistant to combinations of them. [1122] Biofilms composed of *S. au*reus and *C. albicans* are more virulent when co-infecting hosts. [1123] One explanation could be that mixed species biofilms enhance resistance from all the species of which they are composed. [1124, 1125]

Multispecies biofilms have a complex network of interspecific interactions that are poorly understand. Such biofilms are far more complex than single-species biofilms; therefore, principles obtained with single-species biofilms are not directly applicable to multispecies biofilms. Research of biofilms should be focused in a holistic approach, studying interactions between different species and their impact on biofilm resistance. [40]

HYGIENE OF BIOFILMS

Biofilms have multiple impacts on hygiene in food safety and hospital-acquired infections. The control of biofilms is a vital issue because of its complexity, and because resistance to antimicrobials is unavoidable. Researchers should be encouraged to continually explore new strategies and new compounds for biofilm eradication. Moreover, the development of cleaning regimes, product quality and economic issues should be taken in to the account. [1034] A better understanding of biofilm formation process from adhesion to maturation will increase the possibility of efficient control. [51, 244, 1126] In HACCP plans, measures to control contaminants and potentially reduce biocide efficacy should be implemented. Monitoring of biofilms in processing lines and the plant layout has to be assured of an efficient cleaning and disinfection program (Figure 32).

These new areas of investigation will be facilitated by new experimental approaches, analytical techniques and insights, including (but not limited to) in-situ visualization of biofilm development under flux conditions [148], advanced simulation approaches and investigation of biological methods for biofilm prevention and control.

Chapter 11

CONCLUSION

Bacterial biofilms are ubiquitous in our environment. Food industry and hospital facilities are no exception to this. Moreover, biofilms formed on food-processing equipment and medical devices act as a persistent source of contamination, resulting in food-borne disease and hospital-acquired infections. Biofilm prevention and control is, therefore, a priority and should be stimulated to (Figure 32):

- Develop cleaning and disinfection programs that can prevent or eradicate biofilms,
- Select proper contact materials, in addition to their thermal, mechanical, chemical resistance, biocompatibility, as an element of the hygienic design.
- Identify biofilm risk areas in the process and systematically monitor organic and microbial contamination.
- Be aware of factors involved in attachment and biofilm formation and the decreased sensitivity of biofilm to disinfectants.

Pathogenic microorganisms in biofilms represent a source of food contamination. Cleaning and maintenance is one of the most important processes in the food industry; therefore, classic cleaning programs should be upgraded with requirements for the elimination of biofilm. Although conventional control strategies are still used and still being developed, a more economical and environmentally friendly control strategy should satisfy the required hygiene design. Hygiene decision makers should take into account the fact that disinfection methods shall provide a cost-effective result with no adverse effects on human health or to the environment. As published research shows, the hurdle technology has some advantages since it provides synergistic effects, which reduce the material and energy consumption. In addition, inhibiting biofilms and quorum sensing by natural antimicrobials shows some potential for the prevention and control of biofilms.

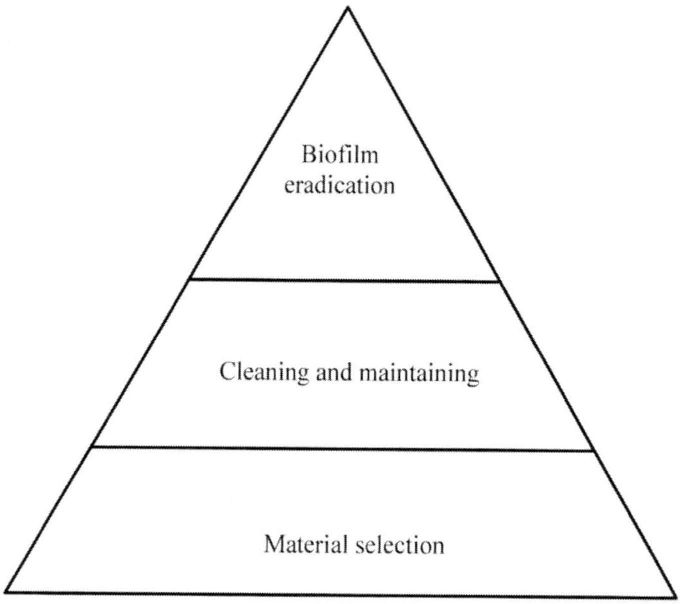

Figure 32. Steps for maintaining surface hygienic.

The growth of biofilms in hospital environments involves multiple physicochemical and biological parameters depending on the properties of the bacteria, the materials, and on the environment. In order to control hospital-acquired infections, strategies should be both efficient against bacteria and should not affect human cells. New strategies, such as surface-modified coating, show some potential in medicine. In the future, questions regarding the impacts of nanostructures on biofilm prevention will have to be answered. The idea of bio-inspired antimicrobial coatings is also an interesting approach that should receive some attention.

The aim of biofilm control is the elimination or reduction of bacterial cells and their activity to an acceptable level. [202] Therefore, programs are established to control microbial proliferation, such as Good Manufacturing Practice and Hazard Analysis and Critical Control Points plans. [44, 70, 1034]

An ineffective cleaning plan leads to operation and maintenance costs caused by the downtime of the production and by contaminated or spoiled products. More importantly, an inadequate cleaning program can be a reason for food poisoning, hospital-acquired infection or even the deaths of patients. The choice of material for surfaces also plays an important role. The risk of biofilm formation is present in almost every operation, unattended risk areas, inconsistent raw water composition, lack of cleaning after failures, and poor access to surfaces. The risk can be reduced by the exclusion of light, use of short piping, inert materials and smooth materials, good air circulation, working at low temperatures, dry conditions [1127], proper handling of implant material, pre-operation skin cleaning and disinfection and post-operative wound treatment. [435, 1107, 1128]

Cleaning and disinfection should be a systematic, planned and controlled action. Knowing the source of contamination is essential in order to define the right antibacterial agent or its combinations, location of application, concentration, temperature levels and exposure time. The type and nature of contamination are also important factors. All the factors abovementioned must be taken into account when designing a disinfection plan. [202] In order to achieve hygiene conditions, actions such as monitoring the presence of biofilm and control efficiency are required.

Research on microbial biofilms is in progress on many fronts, with emphasis on various prevention or control strategies and the development of new approaches for assessing the efficacy of these treatments. Future research should also focus on the role of biofilms in hygiene design. The importance of biofilms is far broader than that of microbiology. Knowledge of material science, chemistry, physics, engineering and some other fields makes vital contributions to understanding the nature of life on surfaces.

REFERENCES

[1] Blackledge MS, Worthington RJ, Melander C. Biologically inspired strategies for combating bacterial biofilms. Current Opinion in Pharmacology 2013; 13: 699-706.

[2] Roberts RR, Hota B, Ahmad L et al. Hospital and societal costs of antimicrobial-resistant infections in a chicago teaching hospital: implications for antibiotic stewardship. *Clinical Infectious Diseases* 2009; 49: 1175-1184.

[3] Davies D. Understanding biofilm resistance to antibacterial agents. *Nature Reviews Drug Discovery* 2003; 2: 114-122.

[4] Van Houdt R, Michiels CW. Biofilm formation and the food industry, a focus on the bacterial outer surface. *Journal of Applied Microbiology* 2010; 109: 1117-1131.

[5] Srey S, Jahid IK, Ha S-D. Biofilm formation in food industries: A food safety concern. *Food Control* 2013; 31: 572-585.

[6] Shi X, Zhu X. Biofilm formation and food safety in food industries. *Trends in Food Science and Technology* 2009; 20: 407-413.

[7] Suleman L, Archer D, Cochrane CA et al. Chapter Ten - Healthcare-Associated Infections and Biofilms. In *Biofilms in Infection Prevention and Control.* Boston: Academic Press; 2014:165-184.

[8] Donlan RM. Biofilms: Microbial life on surfaces. *Emerging Infectious Diseases* 2002; 8: 881-890.

[9] Filip S, Fink R, Oder M et al. Hygienic acceptance of wood in food industry. *Wood Science and Technology* 2012; 46: 657-665.

[10] Simões M, Simões LC, Vieira MJ. A review of current and emergent biofilm control strategies. *LWT - Food Science and Technology* 2010; 43: 573-583.

[11] Costerton JW, Stewart PS, Greenberg EP. Bacterial biofilms: A common cause of persistent infections. *Science* 1999; 284: 1318-1322.

[12] Donlan RM, Costerton JW. Biofilms: Survival mechanisms of clinically relevant microorganisms. *Clinical Microbiology Reviews* 2002; 15: 167-193.

[13] Stoodley P, Lewandowski Z, Boyle JD et al. Structural deformation of bacterial biofilms caused by short-term fluctuations in fluid shear: An in situ investigation of biofilm rheology. *Biotechnology and Bioengineering* 1999; 65: 83-92.

[14] Jahid IK, Ha SD. A review of microbial biofilms of produce: Future challenge to food safety. *Food Science and Biotechnology* 2012; 21: 299-316.

[15] Pratt LA, Kolter R. Genetic analysis of Escherichia coli biofilm formation: Roles of flagella, motility, chemotaxis and type I pili. *Molecular Microbiology* 1998; 30: 285-293.

[16] O'Toole G, Kaplan HB, Kolter R. Biofilm formation as microbial development. *Annual Review of Microbiology 2000*; 54: 49-79.

[17] Prigent-Combaret C, Brombacher E, Vidal O et al. Complex regulatory network controls initial adhesion and biofilm formation in Escherichia coli via regulation of the csgD gene. *Journal of Bacteriology* 2001; 183: 7213-7223.

[18] Parsek MR, Singh PK. Bacterial Biofilms: An Emerging Link to Disease Pathogenesis. *Annual Review of Microbiology* 2003; 57: 677-701.

[19] Lenz AP, Williamson KS, Pitts B et al. Localized gene expression in Pseudomonas aeruginosa biofilms. *Applied and Environmental Microbiology* 2008; 74: 4463-4471.

[20] Monds RD, O'Toole GA. The developmental model of microbial biofilms: ten years of a paradigm up for review. *Trends in Microbiology* 2009; 17: 73-87.

[21] Whiteley M, Bangera MG, Bumgarner RE et al. Gene expression in Pseudomonas aeruginosa biofilms. *Nature* 2001; 413: 860-864.

[22] Schembri MA, Kjærgaard K, Klemm P. Global gene expression in Escherichia coli biofilms. *Molecular Microbiology* 2003; 48: 253-267.

[23] Vuong C, Voyich JM, Fischer ER et al. Polysaccharide intercellular adhesin (PIA) protects Staphylococcus epidermidis against major components of the human innate immune system. *Cellular Microbiology* 2004; 6: 269-275.

[24] Zottola EA, Sasahara KC. Microbial biofilms in the food processing industry - Should they be a concern? *International Journal of Food Microbiology* 1994; 23: 125-148.

[25] Heukelekian H, Heller A. Relation between food concentration and surface for bacterial growth. *J Bacteriol* 1940; 40: 547-558.

[26] Zobell CE. The effect of solid surfaces upon bacterial activity. *J Bacteriol* 1943; 46: 39-56.

[27] Jones HC, Roth IL, Sanders 3rd WM. Electron microscopic study of a slime layer. *Journal of Bacteriology* 1969; 99: 316-325.

[28] Characklis WG. Attached microbial growths. II. Frictional resistance due to microbial slimes. *Water Research* 1973; 7: 1249-1258.

[29] Costerton JW, Geesey GG, Cheng KJ. *How bacteria stick. Scientific American* 1978; 238: 86-95.

[30] Shanmugasundaram N, Sundaraseelan J, Uma S et al. Design and delivery of silver sulfadiazine from alginate microspheres-impregnated collagen scaffold. *Journal of Biomedical Materials Research - Part B Applied Biomaterials* 2006; 77: 378-388.

[31] Watnick P, Kolter R. Biofilm, city of microbes. *Journal of Bacteriology* 2000; 182: 2675-2679.

[32] Mah TFC, O'Toole GA. Mechanisms of biofilm resistance to antimicrobial agents. *Trends in Microbiology* 2001; 9: 34-39.

[33] Anderl JN, Franklin MJ, Stewart PS. Role of antibiotic penetration limitation in Klebsiella pneumoniae biofilm resistance to ampicillin and ciprofloxacin. *Antimicrobial Agents and Chemotherapy* 2000; 44: 1818-1824.

[34] Zhang L, Mah TF. Involvement of a novel efflux system in biofilm-specific resistance to antibiotics. *Journal of Bacteriology* 2008; 190: 4447-4452.

[35] Worthington RJ, Blackledge MS, Melander C. Small-molecule inhibition of bacterial two-component systems to combat antibiotic resistance and virulence. *Future Medicinal Chemistry* 2013; 5: 1265-1284.

[36] Sun F, Qu F, Ling Y et al. Biofilm-associated infections: Antibiotic resistance and novel therapeutic strategies. *Future Microbiology* 2013; 8: 877-886.

[37] Tan SYE, Chew SC, Tan SYY et al. Emerging frontiers in detection and control of bacterial biofilms. *Current Opinion in Biotechnology* 2014; 26: 1-6.

[38] Sanderson AR, Leid JG, Hunsaker D. Bacterial biofilms on the sinus mucosa of human subjects with chronic rhinosinusitis. *Laryngoscope* 2006; 116: 1121-1126.

[39] Mitri S, Xavier JB, Foster KR. Social evolution in multispecies biofilms. *Proceedings of the National Academy of Sciences of the United States of America* 2011; 108: 10839-10846.

[40] Burmølle M, Ren D, Bjarnsholt T et al. Interactions in multispecies biofilms: do they actually matter? *Trends in Microbiology* 2014; 22: 84-91.

[41] Ukuku DO, Sapers GM. Effect of sanitizer treatments on Salmonella Stanley attached to the surface of cantaloupe and cell transfer to fresh-cut tissues during cutting practices. *Journal of Food Protection* 2001; 64: 1286-1291.

[42] Kreske AC, Ryu JH, Pettigrew CA et al. Lethality of chlorine, chlorine dioxide, and a commercial produce sanitizer to Bacillus cereus and Pseudomonas in a liquid detergent, on stainless steel, and in biofilm. *Journal of Food Protection* 2006; 69: 2621-2634.

[43] Keskinen LA, Annous BA. Efficacy of adding detergents to sanitizer solutions for inactivation of Escherichia coli O157:H7 on Romaine lettuce. *International Journal of Food* Microbiology 2011; 147: 157-161.

[44] Sharma M, Anand SK. Biofilms evaluation as an essential component of HACCP for food/dairy processing industry - A case. *Food Control* 2002; 13: 469-477.

[45] Stubblefield BA, Howery KE, Islam BN et al. Constructing multispecies biofilms with defined compositions by sequential deposition of bacteria. *Applied Microbiology and* Biotechnology 2010; 86: 1941-1946.

[46] Møretrø T, Langsrud S. Listeria monocytogenes: Biofilm formation and persistence in food-processing environments. *Biofilms* 2004; 1: 107-121.

[47] Fux CA, Costerton JW, Stewart PS et al. Survival strategies of infectious biofilms. *Trends in Microbiology* 2005; 13: 34-40.

[48] Xu KD, McFeters GA, Stewart PS. Biofilm resistance to antimicrobial agents. *Microbiology* 2000; 146: 547-549.

[49] Van Der Veen S, Abee T. Importance of SigB for listeria monocytogenes static and continuous-flow biofilm formation and disinfectant resistance. *Applied and Environmental Microbiology* 2010; 76: 7854-7860.

[50] McLandsborough L, Rodriguez A, Pérez-Conesa D et al. Biofilms: At the interface between biophysics and microbiology. *Food Biophysics* 2006; 1: 94-114.

[51] Simões LC, Lemos M, Pereira AM et al. Persister cells in a biofilm treated with a biocide. *Biofouling* 2011; 27: 403-411.

[52] Noimark S, Dunnill CW, Wilson M et al. The role of surfaces in catheter-associated infections. *Chemical Society Reviews* 2009; 38: 3435-3448.

[53] Nakamura H, Takakura KI, Sone Y et al. Biofilm formation and resistance to benzalkonium chloride in listeria monocytogenes isolated from a fish processing plant. *Journal of Food Protection* 2013; 76: 1179-1186.

[54] Campoccia D, Montanaro L, Arciola CR. A review of the biomaterials technologies for infection-resistant surfaces. *Biomaterials* 2013; 34: 8533-8554.

[55] Campoccia D, Montanaro L, Arciola CR. The significance of infection related to orthopedic devices and issues of antibiotic resistance. *Biomaterials* 2006; 27: 2331-2339.

[56] Montanaro L, Speziale P, Campoccia D et al. Scenery of Staphylococcus implant infections in orthopedics. *Future Microbiology* 2011; 6: 1329-1349.

[57] Wenzel RP. Health care-associated infections: Major issues in the early years of the 21st century. *Clinical Infectious Diseases* 2007; 45: S85-S88.

[58] Bryers JD. Medical biofilms. *Biotechnology and Bioengineering* 2008; 100: 1-18.

[59] Wirtanen G. Food process hygiene, effective cleaning and safety in the food industry. *Microbial Contaminants and Contamination Routes in Food Industry* 2007: 129-144.

[60] Theyssen H, Stanga M, Bruschi F et al. Sanitizing and Cleaning composition and its use for Sanitizing and/or Cleaning hard surfaces. *US Patent* 2006; 503: 488.

[61] Neu TR, Manz B, Volke F et al. Advanced imaging techniques for assessment of structure, composition and function in biofilm systems. *FEMS Microbiology Ecology* 2010; 72: 1-21.

[62] Stoodley P, Boyle JD, Dodds I et al. Consensus model of biofilm structure. *Biofilms: Community Interactions and Control* 1997: 1-9.

[63] Vieira MJ, Melo LF, Pinheiro MM. Biofilm formation: Hydrodynamic effects on internal diffusion and structure. *Biofouling* 1993; 7: 67-80.

[64] Gilbert P, McBain AJ. Potential impact of increased use of biocides in consumer products on prevalence of antibiotic resistance. *Clinical Microbiology Reviews* 2003; 16: 189-208.

[65] Momba MNB, Cloete TE, Venter SN et al. Evaluation of the impact of disinfection processes on the formation of biofilms in potable surface water distribution systems. *Water Science and Technology* 1998; 38: 283-289.

[66] MacDonald A, Dinah F, MacKenzie D et al. Performance feedback of hand hygiene, using alcohol gel as the skin decontaminant, reduces the number of inpatients newly affected by MRSA and antibiotic costs. *Journal of Hospital Infection* 2004; 56: 56-63.

[67] Cloete TE, Jacobs L, Brözel VS. The chemical control of biofouling in industrial water systems. *Biodegradation* 1998; 9: 23-37.

[68] Ferreira C, Pereira AM, Melo LF et al. Advances in industrial biofilm control with micro-nanotechnology. *Current Research, Technology and Education Topics in Applied Microbiology and Microbial Biotechnology* 2010; 2: 845-854.

[69] Hygienic Equipment Design Criteria. In. Frankfurt, Germany: E*uropean Hygienic Engineering & Design Group;* 2004:16.

[70] Gibson H, Taylor JH, Hall KE et al. Effectiveness of cleaning techniques used in the food industry in terms of the removal of bacterial biofilms. *Journal of Applied Microbiology* 1999; 87: 41-48.

[71] Bixler GD, Bhushan B. Review article: Biofouling: Lessons from nature. *Philosophical Transactions of the Royal Society A: Mathematical, Physical and Engineering Sciences* 2012; 370: 2381-2417.

[72] Bohinc K, Dražić G, Fink R et al. Available surface dictates microbial adhesion capacity. *International Journal of Adhesion and Adhesives* 2014; 50: 265-272.

[73] Zhao Q, Wang C, Liu Y et al. Bacterial adhesion on the metal-polymer composite coatings. *International Journal of Adhesion and Adhesives* 2007; 27: 85-91.

[74] Habimana O, Semião AJC, Casey E. The role of cell-surface interactions in bacterial initial adhesion and consequent biofilm formation on nanofiltration/reverse osmosis membranes. *Journal of Membrane Science* 2014; 454: 82-96.

[75] Marshall KC, Stout R, Mitchell R. Mechanism of the initial events in the sorption of marine bacteria to surfaces. *J Gen Microbiol* 1971; 68: 337-348.

[76] Abu-Lail NI, Camesano TA. Role of ionic strength on the relationship of biopolymer conformation, DLVO contributions, and steric interactions to bioadhesion of pseudomonas putida KT2442. *Biomacromolecules* 2003; 4: 1000-1012.

[77] Bolster CH, Mills AL, Hornberger GM et al. Effect of surface coatings, grain size, and ionic strength on the maximum attainable coverage of bacteria on sand surfaces. *Journal of Contaminant Hydrology* 2001; 50: 287-305.

[78] Zhao W, Walker SL, Huang Q et al. Adhesion of bacterial pathogens to soil colloidal particles: Influences of cell type, natural organic matter, and solution chemistry. *Water Research* 2014; 53: 35-46.

[79] Gordesli FP, Abu-Lail NI. Combined poisson and soft-particle DLVO analysis of the specific and nonspecific adhesion forces measured between L. monocytogenes grown at various temperatures and silicon nitride. *Environmental Science and Technology* 2012; 46: 10089-10098.

[80] Bos R, Van Der Mei HC, Busscher HJ. Physico-chemistry of initial microbial adhesive interactions - Its mechanisms and methods for study. *FEMS Microbiology Reviews* 1999; 23: 179-229.

[81] Poortinga AT, Bos R, Norde W et al. Electric double layer interactions in bacterial adhesion to surfaces. *Surface Science Reports* 2002; 47: 1-32.

[82] Hori K, Matsumoto S. Bacterial adhesion: From mechanism to control. *Biochemical Engineering* Journal 2010; 48: 424-434.

[83] Van Oss CJ, Chaudhury MK, Good RJ. Interfacial Lifshitz-van der Waals and polar interactions in macroscopic systems. *Chemical Reviews* 1988; 88: 927-941.

[84] Mukherjee J, Karunakaran E, Biggs CA. Using a multi-faceted approach to determine the changes in bacterial cell surface properties influenced by a biofilm lifestyle. *Biofouling* 2012; 28: 1-14.

[85] Long G, Zhu P, Shen Y et al. Influence of extracellular polymeric substances (EPS) on deposition kinetics of bacteria. *Environmental Science and Technology* 2009; 43: 2308-2314.

[86] Foka A, Katsikogianni MG, Anastassiou ED et al. The combined effect of surface chemistry and flow conditions on Staphylococcus epidermidis adhesion and ica operon expression. *European Cells and Materials* 2012; 24: 386-402.

[87] Monds RD, Newell PD, Gross RH et al. Phosphate-dependent modulation of c-di-GMP levels regulates Pseudomonas fluorescens Pf0-

1 biofilm formation by controlling secretion of the adhesin LapA. *Molecular Microbiology* 2007; 63: 656-679.

[88] Newell PD, Boyd CD, Sondermann H et al. A c-di-GMP effector system controls cell adhesion by inside-out signaling and surface protein cleavage. *PLoS Biology* 2011; 9.

[89] Vrouwenvelder JS, Beyer F, Dahmani K et al. Phosphate limitation to control biofouling. *Water Research* 2010; 44: 3454-3466.

[90] Beveridge TJ. Ultrastructure, chemistry, and function of the bacterial wall. *International Review of Cytology* 1981; Vol.72: 229-317.

[91] Delcour J, Ferain T, Deghorain M et al. The biosynthesis and functionality of the cell-wall of lactic acid bacteria. Antonie van Leeuwenhoek, *International Journal of General and Molecular Microbiology* 1999; 76: 159-184.

[92] Beveridge TJ. Structures of gram-negative cell walls and their derived membrane vesicles. *Journal of Bacteriology* 1999; 181: 4725-4733.

[93] Millsap KW, Reid G, Van Der Mei HC et al. Cluster analysis of genotypically characterized Lactobacillus species based on physicochemical cell surface properties and their relationship with adhesion to hexadecane. *Canadian Journal of Microbiology* 1997; 43: 284-291.

[94] Reid G, Bialkowska-Hobrzanska H, Van Der Mei HC et al. Correlation between genetic, physico-chemical surface characteristics and adhesion of four strains of Lactobacillus. *Colloids and Surfaces B: Biointerfaces* 1999; 13: 75-81.

[95] Geoghegan M, Andrews JS, Biggs CA et al. The polymer physics and chemistry of microbial cell attachment and adhesion. *Faraday Discussions* 2008; 139: 85-103.

[96] An YH, Friedman RJ. Concise review of mechanisms of bacterial adhesion to biomaterial surfaces. *Journal of Biomedical Materials Research* 1998; 43: 338-348.

[97] Fletcher M. Bacterial attachment in aquatic environments: A diversity of surfaces and adhesion strategies. *Bacterial Adhesion: Molecular and Ecological Diversity* 1996: 1-24.

[98] van Loosdrecht MC, Lyklema J, Norde W et al. The role of bacterial cell wall hydrophobicity in adhesion. *Applied and Environmental Microbiology* 1987; 53: 1893-1897.

[99] Pang CM, Hong P, Guo H et al. Biofilm formation characteristics of bacterial isolates retrieved from a reverse osmosis membrane. *Environmental Science and Technology* 2005; 39: 7541-7550.

[100] Herzberg M, Kang S, Elimelech M. Role of extracellular polymeric substances (EPS) in biofouling of reverse osmosis membranes. *Environmental Science and Technology* 2009; 43: 4393-4398.

[101] Habimana O, Meyrand M, Meylheuc T et al. Genetic features of resident biofilms determine attachment of Listeria monocytogenes. *Applied and Environmental Microbiology* 2009; 75: 7814-7821.

[102] Berk V, Fong JCN, Dempsey GT et al. Molecular architecture and assembly principles of Vibrio cholerae biofilms. *Science* 2012; 337: 236-239.

[103] Biesheuvel PM, Van Der Veen M, Norde W. A modified Poisson-Boltzmann model including charge regulation for the adsorption of ionizable polyelectrolytes to charged interfaces, applied to lysozyme adsorption on silica. *Journal of Physical Chemistry* B 2005; 109: 4172-4180.

[104] Pesce G, Rusciano G, Sasso A et al. Surface charge and hydrodynamic coefficient measurements of Bacillus subtilis spore by optical tweezers. *Colloids and Surfaces B: Biointerfaces* 2014; 116: 568-575.

[105] Subramani A, Hoek EMV. Direct observation of initial microbial deposition onto reverse osmosis and nanofiltration membranes. *Journal of Membrane Science* 2008; 319: 111-125.

[106] Van Merode AEJ, Van Der Mei HC, Busscher HJ et al. Influence of culture heterogeneity in cell surface charge on adhesion and biofilm formation by Enterococcus faecalis. *Journal of Bacteriology* 2006; 188: 2421-2426.

[107] Van Merode AEJ, van der Mei HC, Busscher HJ et al. Enterococcus faecalis strains show culture heterogeneity in cell surface charge. *Microbiology* 2006; 152: 807-814.

[108] Hong Y, Brown DG. Electrostatic behavior of the charge-regulated bacterial cell surface. *Langmuir* 2008; 24: 5003-5009.

[109] Bunt CR, Jones DS, Tucker IG. The effects of pH, ionic strength and polyvalent ions on the cell surface hydrophobicity of Escherichia coli evaluated by the BATH and HIC methods. *International Journal of Pharmaceutics* 1995; 113: 257-261.

[110] Song B, Leff LG. Influence of magnesium ions on biofilm formation by Pseudomonas fluorescens. *Microbiological Research* 2006; 161: 355-361.

[111] Zulfakar SS, White JD, Ross T et al. Effect of pH, salt and chemical rinses on bacterial attachment to extracellular matrix proteins. *Food Microbiology* 2013; 34: 369-375.

[112] Lim Y, Jana M, Luong TT et al. Control of Glucose- and NaCl-Induced Biofilm Formation by rbf in Staphylococcus aureus. *Journal of Bacteriology* 2004; 186: 722-729.

[113] Liu ZD, Li JY, Jiang J et al. Adhesion of Escherichia coli to nano-Fe/Al oxides and its effect on the surface chemical properties of Fe/Al oxides. *Colloids and Surfaces B: Biointerfaces* 2013; 110: 289-295.

[114] Chen G, Walker SL. Role of solution chemistry and ion valence on the adhesion kinetics of groundwater and marine bacteria. *Langmuir* 2007; 23: 7162-7169.

[115] Chen G. Escherichia coli adhesion to abiotic surfaces in the presence of non-ionic surfactants. *Journal of Adhesion Science and Technology* 2003; 17: 2131-2139.

[116] Van Hoogmoed CG, Van Der Mei HC, Busscher HJ. The influence of calcium on the initial adhesion of S. thermophilus to stainless steel under flow studied by metallurgical microscopy. *Biofouling* 1997; 11: 167-176.

[117] Rzhepishevska O, Hakobyan S, Ruhal R et al. The surface charge of anti-bacterial coatings alters motility and biofilm architecture. Biomaterials Science 2013; 1: 589-602.

[118] Kim MG, Oh MH, Lee GY et al. Analysis of major foodborne pathogens in various foods in Korea. *Food Science and Biotechnology* 2008; 17: 483-488.

[119] Lemon KP, Higgins DE, Kolter R. Flagellar motility is critical for Listeria monocytogenes biofilm formation. *Journal of Bacteriology* 2007; 189: 4418-4424.

[120] Vatanyoopaisarn S, Nazli A, Dodd CER et al. Effect of flagella on initial attachment of Listeria monocytogenes to stainless steel. *Applied and Environmental Microbiology* 2000; 66: 860-863.

[121] DeFlaun MF, Tanzer AS, McAteer AL et al. Development of an adhesion assay and characterization of an adhesion-deficient mutant of Pseudomonas fluorescens. *Applied and Environmental Microbiology* 1990; 56: 112-119.

[122] Low D, Braaten B, Van Der Woude M. Fimbriae. *Escherichia Coli and Salmonella* 1996: 146-157.

[123] Klemm P, Krogfelt KA. Type 1 fimbriae of Escherichia coli. *Fimbriae, Adhesion, Genetics, Biogenesis and Vaccines 1994*: 9-26.

[124] Ren D, Bedzyk LA, Thomas SM et al. Gene expression in Escherichia coli biofilms. *Applied Microbiology and Biotechnology* 2004; 64: 515-524.

[125] Liu NT, Nou X, Lefcourt AM et al. Dual-species biofilm formation by Escherichia coli O157:H7 and environmental bacteria isolated from fresh-cut processing facilities. *International Journal of Food Microbiology* 2014; 171: 15-20.

[126] Król JE, Wojtowicz AJ, Rogers LM et al. Invasion of E. coli biofilms by antibiotic resistance plasmids. *Plasmid* 2013; 70: 110-119.

[127] Cookson AL, Cooley WA, Woodward MJ. The role of type 1 and curli fimbriae of Shiga toxin-producing Escherichia coli in adherence to abiotic surfaces. *International Journal of Medical Microbiology* 2002; 292: 195-205.

[128] Di Martino P, Cafferini N, Joly B et al. Klebsiella pneumoniae type 3 pili facilitate adherence and biofilm formation on abiotic surfaces. *Research in Microbiology* 2003; 154: 9-16.

[129] Béchet M, Blondeau R. Factors associated with the adherence and biofilm formation by Aeromonas caviae on glass surfaces. *Journal of Applied Microbiology* 2003; 94: 1072-1078.

[130] Ryu JH, Kim H, Frank JF et al. Attachment and biofilm formation on stainless steel by Escherichia coli O157:H7 as affected by curli production. *Letters in Applied Microbiology* 2004; 39: 359-362.

[131] Telford JL, Barocchi MA, Margarit I et al. Pili in Gram-positive pathogens. *Nature Reviews Microbiology* 2006; 4: 509-519.

[132] Ton-That H, Marraffini LA, Schneewind O. Protein sorting to the cell wall envelope of Gram-positive bacteria. *Biochimica et Biophysica Acta - Molecular Cell Research* 2004; 1694: 269-278.

[133] Barak JD, Jahn CE, Gibson DL et al. The role of cellulose and O-antigen capsule in the colonization of plants by Salmonella enterica. *Molecular Plant-Microbe Interactions* 2007; 20: 1083-1091.

[134] Genevaux P, Bauda P, DuBow MS et al. Identification of Tn 10 insertions in the rlaG, rfaP, and gaLU genes involved in lipopolysaccharide core biosynthesis that affect Escherichia coli adhesion. *Archives of Microbiology* 1999; 172: 1-8.

[135] Monk IR, Cook GM, Monk BC et al. Morphotypic conversion in Listeria monocytogenes biofilm formation: Biological significance of rough colony isolates. *Applied and Environmental Microbiology* 2004; 70: 6686-6694.

[136] Jass J, Surman S, Walker J. Microbial biofilms in medicine. *Medical Biofilms: Detection, Prevention and Control* 2003: 1-28.

[137] Walker SL, Redman JA, Elimelech M. Role of cell surface lipopolysaccharides in escherichia coli K12 adhesion and transport. *Langmuir* 2004; 20: 7736-7746.

[138] Perni S, Preedy EC, Prokopovich P. Success and failure of colloidal approaches in adhesion of microorganisms to surfaces. *Advances in Colloid and Interface Science* 2013;.

[139] Busscher HJ, Rinastiti M, Siswomihardjo W et al. Biofilm formation on dental restorative and implant materials. *Journal of Dental Research* 2010; 89: 657-665.

[140] Bendersky M, Davis JM. DLVO interaction of colloidal particles with topographically and chemically heterogeneous surfaces. *Journal of Colloid and Interface Science* 2011; 353: 87-97.

[141] Liu LF, Yu SC, Wu LG et al. Study on a novel polyamide-urea reverse osmosis composite membrane (ICIC-MPD). II. Analysis of membrane antifouling performance. *Journal of Membrane Science* 2006; 283: 133-146.

[142] Cunliffe D, Smart CA, Alexander C et al. Bacterial adhesion at synthetic surfaces. *Applied and Environmental Microbiology* 1999; 65: 4995-5002.

[143] Faille C, Jullien C, Fontaine F et al. Adhesion of Bacillus spores and Escherichia coli cells to inert surfaces: Role of surface hydrophobicity. *Canadian Journal of Microbiology* 2002; 48: 728-738.

[144] Campbell P, Srinivasan R, Knoell T et al. Quantitative structure-activity relationship (QSAR) analysis of surfactants influencing attachment of a Mycobacterium sp. to cellulose acetate and aromatic polyamide reverse osmosis membranes. *Biotechnology and Bioengineering* 1999; 64: 527-544.

[145] Kochkodan V, Tsarenko S, Potapchenko N et al. Adhesion of microorganisms to polymer membranes: a photobactericidal effect of surface treatment with TiO2. *Desalination* 2008; 220: 380-385.

[146] Lee RW, Glater J, Cohen Y et al. Low-pressure RO membrane desalination of agricultural drainage water. *Desalination* 2003; 155: 109-120.

[147] Flemming HC. Reverse osmosis membrane biofouling. *Experimental Thermal and Fluid Science* 1997; 14: 382-391.

[148] Subramani A, Huang X, Hoek EMV. Direct observation of bacterial deposition onto clean and organic-fouled polyamide membranes. *Journal of Colloid and Interface Science* 2009; 336: 13-20.

[149] Myint AA, Lee W, Mun S et al. Influence of membrane surface properties on the behavior of initial bacterial adhesion and biofilm development onto nanofiltration membranes. *Biofouling* 2010; 26: 313-321.

[150] Dorobantu LS, Gray MR. Application of atomic force microscopy in bacterial research. *Scanning* 2010; 32: 74-96.

[151] Crawford RJ, Webb HK, Truong VK et al. Surface topographical factors influencing bacterial attachment. *Advances in Colloid and Interface Science* 2012; 179-182: 142-149.

[152] Donlan R, Murga R, Carpenter J et al. Monochloramine disinfection of biofilm-associated Legionella pneumophila in a potable water model system. *Legionella* 2002: 406-410.

[153] Whitehead KA, Verran J. The effect of surface topography on the retention of microorganisms. *Food and Bioproducts Processing* 2006; 84: 253-259.

[154] Wu Y, Zitelli JP, TenHuisen KS et al. Differential response of Staphylococci and osteoblasts to varying titanium surface roughness. *Biomaterials* 2011; 32: 951-960.

[155] Taylor RL, Verran J, Lees GC et al. The influence of substratum topography on bacterial adhesion to polymethyl methacrylate. *Journal of Materials Science: Materials in Medicine* 1998; 9: 17-22.

[156] Zhou Z, Wu P, Ma C. Hydrophobic interaction and stability of collidal silica. *Colloids and Surfaces* 1990; 50: 177-188.

[157] Bazaka K, Crawford RJ, Ivanova EP. Do bacteria differentiate between degrees of nanoscale surface roughness? *Biotechnology Journal* 2011; 6: 1103-1114.

[158] Truong VK, Lapovok R, Estrin YS et al. The influence of nano-scale surface roughness on bacterial adhesion to ultrafine-grained titanium. *Biomaterials* 2010; 31: 3674-3683.

[159] Ivanova EP, Truong VK, Wang JY et al. Impact of nanoscale roughness of titanium thin film surfaces on bacterial Retention. *Langmuir* 2010; 26: 1973-1982.

[160] Mitik-Dineva N, Wang J, Truong VK et al. Escherichia coli, Pseudomonas aeruginosa, and Staphylococcus aureus Attachment Patterns on Glass Surfaces with Nanoscale Roughness. *Current Microbiology* 2009; 58: 268-273.

[161] Whitehead KA, Colligon J, Verran J. Retention of microbial cells in substratum surface features of micrometer and sub-micrometer dimensions. *Colloids and Surfaces B: Biointerfaces* 2005; 41: 129-138.

[162] Bakker DP, Busscher HJ, van Zanten J et al. Multiple linear regression analysis of bacterial deposition to polyurethane coatings after conditioning film formation in the marine environment. *Microbiology* 2004; 150: 1779-1784.
[163] Giraldez MJ, Resua CG, Lira M et al. Contact lens hydrophobicity and roughness effects on bacterial adhesion. *Optometry and Vision Science* 2010; 87: E426-E431.
[164] Tang H, Cao T, Liang X et al. Influence of silicone surface roughness and hydrophobicity on adhesion and colonization of Staphylococcus epidermidis. *J Biomed Mater Res* A; 88: 454-463.
[165] Singh AV, Vyas V, Patil R et al. Quantitative characterization of the influence of the nanoscale morphology of nanostructured surfaces on bacterial adhesion and biofilm formation. *PLoS One* 2011; 6: 1-12.
[166] Fan PL, Batchu H, Chou HN et al. Laboratory evaluation of amalgam separators. *Journal of the American Dental Association* 2002; 133: 577-584.
[167] Quirynen M, Bollen CM. The influence of surface roughness and surface-free energy on supra- and subgingival plaque formation in man. A review of the literature. *Journal of Clinical Periodontology* 1995; 22: 1-14.
[168] Mei L, Busscher HJ, Van Der Mei HC et al. Influence of surface roughness on streptococcal adhesion forces to composite resins. *Dental Materials* 2011; 27: 770-778.
[169] Kouider N, Hamadi F, Mallouki B et al. Effect of stainless steel surface roughness on Staphylococcus aureus adhesion. *Int J App Sci* 2010; 4: 1-7.
[170] Li B, Logan BE. Bacterial adhesion to glass and metal-oxide surfaces. *Colloids and Surfaces B: Biointerfaces* 2004; 36: 81-90.
[171] Memarzadeh K, Sharili AS, Huang J et al. Nanoparticulate zinc oxide as a coating material for orthopedic and dental implants. *Journal of Biomedical Materials Research - Part* A 2014;.
[172] Gu H, Ren D. Materials and surface engineering to control bacterial adhesion and biofilm formation: A review of recent advances. *Frontiers of Chemical Science and Engineering* 2014; 8: 20-33.
[173] Carvalho I, Henriques M, Oliveira JC et al. Influence of surface features on the adhesion of staphylococcus epidermidis to Ag-TiCN thin films. *Science and Technology of Advanced Materials* 2013; 14.

[174] Bergmans L, Moisiadis P, Van Meerbeek B et al. Microscopic observation of bacteria: Review highlighting the use of environmental SEM. *International Endodontic Journal* 2005; 38: 775-788.

[175] Wittmaack K, Wehnes H, Heinzmann U et al. An overview on bioaerosols viewed by scanning electron microscopy. *Science of the Total Environment* 2005; 346: 244-255.

[176] Merritt K, Hitchins VM, Neale AR. Tissue colonization from implantable biomaterials with low numbers of bacteria. *Journal of Biomedical Materials Research* 1999; 44: 261-265.

[177] Cheng Y, Fan W, Guo L. Coking wastewater treatment using a magnetic porous ceramsite carrier. *Separation and Purification Technology* 2014; 130: 167-172.

[178] Gimeno M, Pinczowski P, Vázquez FJ et al. Porous orthopedic steel implant as an antibiotic eluting device: Prevention of post-surgical infection on an ovine model. *International Journal of Pharmaceutics* 2013; 452: 166-172.

[179] Perez LM, Lalueza P, Monzon M et al. Hollow porous implants filled with mesoporous silica particles as a two-stage antibiotic-eluting device. *International Journal of Pharmaceutics* 2011; 409: 1-8.

[180] Peters G, Locci R, Pulverer G. Adherence and growth of coagulase negative staphylococci on surfaces of intravenous catheters. *Journal of Infectious Diseases* 1982; 146: 479-482.

[181] Mosadegh-Sedghi S, Rodrigue D, Brisson J et al. Wetting phenomenon in membrane contactors - Causes and prevention. *Journal of Membrane Science* 2014; 452: 332-353.

[182] Levin Y, Dos Santos AP. Ions at hydrophobic interfaces. *Journal of Physics Condensed Matter* 2014; 26.

[183] Van Der Mei HC, Bos R, Busscher HJ. A reference guide to microbial cell surface hydrophobicity based on contact angles. *Colloids and Surfaces B: Biointerfaces* 1998; 11: 213-221.

[184] Chavant P, Martinie B, Meylheuc T et al. Listeria monocytogenes LO28: Surface physicochemical properties and ability to form biofilms at different temperatures and growth phases. *Applied and Environmental Microbiology* 2002; 68: 728-737.

[185] Popovici J, White CP, Hoelle J et al. Characterization of the cell surface properties of drinking water pathogens by microbial adhesion to hydrocarbon and electrophoretic mobility measurements. *Colloids and Surfaces B: Biointerfaces* 2014; 118: 126-132.

[186] Fernandes P, São José JFB, Zerdas ERMA et al. Influence of the hydrophobicity and surface roughness of mangoes and tomatoes on the adhesion of Salmonella enterica serovar Typhimurium and evaluation of cleaning procedures using surfactin. *Food Control* 2014; 41: 21-26.

[187] Fletcher M, Lessmann JM, Loeb GI. Bacterial surface adhesives and biofilm matrix polymers of marine and freshwater bacteria. *Biofouling* 1991; 4: 129-140.

[188] Satou N, Satou J, Shintani H et al. Adherence of Streptococci to surface-modified glass. *Journal of General Microbiology* 1988; 134: 1299-1305.

[189] Al-Makhlafi H, McGuire J, Daeschel M. Influence of preadsorbed milk proteins on adhesion of Listeria monocytogenes to hydrophobic and hydrophilic silica surfaces. *Applied and Environmental Microbiology* 1994; 60: 3560-3565.

[190] Bridgett MJ, Davies MC, Denyer SP. Control of staphylococcal adhesion to polystyrene surfaces by polymer surface modification with surfactants. *Biomaterials* 1992; 13: 411-416.

[191] Frueh J, Gai M, Yang Z et al. Influence of polyelectrolyte multilayer coating on the degree and type of biofouling in freshwater environment. *Journal of Nanoscience and Nanotechnology* 2014; 14: 4341-4350.

[192] Saha N, Monge C, Dulong V et al. Influence of polyelectrolyte film stiffness on bacterial growth. *Biomacromolecules* 2013; 14: 520-528.

[193] Graisuwan W, Wiarachai O, Ananthanawat C et al. Multilayer film assembled from charged derivatives of chitosan: Physical characteristics and biological responses. *Journal of Colloid and Interface Science* 2012; 376: 177-188.

[194] Lawrence NJ, Wells-Kingsbury JM, Ihrig MM et al. Controlling E. coli adhesion on high-k dielectric bioceramics films using poly(amino acid) multilayers. *Langmuir* 2012; 28: 4301-4308.

[195] Loeb GI, Neihof RA. Marine conditioning films. *Applied Chemistry at Protein Interfaces* 1975; 145: 319-335.

[196] Marsh PD. Dental plaque. *Microbial Biofilms* 1995: 282-300.

[197] Mittelman MW. Adhesion to biomaterials. *Bacterial Adhesion: Molecular and Ecological Diversity* 1996: 89-127.

[198] Ofek I, Doyle RJ. *Bacterial Adhesion to Cells and Tissues* 1994;.

[199] Stoodley P, Sauer K, Davies DG et al. Biofilms as complex differentiated communities. *Annual Review of Microbiology* 2002; 56: 187-209.

[200] Gerke C, Kraft A, Süßmuth R et al. Characterization of the N-Acetylglucosaminyltransferase activity involved in the biosynthesis of the Staphylococcus epidermidis polysaccharide intercellular adhesin. *Journal of Biological Chemistry* 1998; 273: 18586-18593.

[201] Sinde E, Carballo J. Attachment of Salmonella spp. and Listeria monocytogenes to stainless steel, rubber and polytetrafluorehtylene: The influence of free energy and the effect of commercial sanitizers. *Food Microbiology* 2000; 17: 439-447.

[202] Maukonen J, Mättö J, Wirtanen G et al. Methodologies for the characterization of microbes in industrial environments: A review. *Journal of Industrial Microbiology and Biotechnology* 2003; 30: 327-356.

[203] Chae MS, Schraft H. Cell viability of Listeria monocytogenes biofilms. *Food Microbiology* 2001; 18: 103-112.

[204] Oulahal N, Brice W, Martial A et al. Quantitative analysis of survival of Staphylococcus aureus or Listeria innocua on two types of surfaces: Polypropylene and stainless steel in contact with three different dairy products. *Food Control* 2008; 19: 178-185.

[205] Patel JD, Ebert M, Ward R et al. S. epidermidis biofilm formation: Effects of biomaterial surface chemistry and serum proteins. *Journal of Biomedical Materials Research* - Part A 2007; 80: 742-751.

[206] Simões M, Simões LC, Cleto S et al. The effects of a biocide and a surfactant on the detachment of Pseudomonas fluorescens from glass surfaces. *International Journal of Food Microbiology* 2008; 121: 335-341.

[207] Busscher HJ, Van Der Mei HC. Microbial adhesion in flow displacement systems. *Clinical Microbiology Reviews 2006*; 19: 127-141.

[208] Allison DG. The biofilm matrix. *Biofouling* 2003; 19: 139-150.

[209] Parsek MR, Greenberg EP. Sociomicrobiology: The connections between quorum sensing and biofilms. *Trends in Microbiology* 2005; 13: 27-33.

[210] Sauer K, Camper AK. Characterization of phenotypic changes in Pseudomonas putida in response to surface-associated growth. *Journal of Bacteriology* 2001; 183: 6579-6589.

[211] Gristina AG, Hobgood CD, Webb LX et al. Adhesive colonization of biomaterials and antibiotic resistance. *Biomaterials* 1987; 8. 423-426.

[212] Brading MG, Jass J, Lappin-Scott HM. Dynamics of bacterial biofilm formation. *Microbial Biofilms* 1995: 46-63.

[213] Schneider RH, Nidich SI, Salerno JW et al. Lower lipid peroxide levels in practitioners of the transcendental meditation® program. *Psychosomatic Medicine* 1998; 60: 38-41.

[214] Atabek A, Camesano TA. Atomic force microscopy study of the effect of lipopolysaccharides and extracellular polymers on adhesion of Pseudomonas aeruginosa. *Journal of Bacteriology* 2007; 189: 8503-8509.

[215] Marshall KC, Cruickshank RH. Cell surface hydrophobicity and the orientation of certain bacteria at interfaces. *Archiv für Mikrobiologie* 1973; 91: 29-40.

[216] Doyle RJ. *Microbial Growth in Biofilms* - Part A: Developmental and Molecular Biological Aspects 2001;.

[217] Sutherland IW. The biofilm matrix - An immobilized but dynamic microbial environment. *Trends in Microbiology* 2001; 9: 222-227.

[218] Azeredo J, Oliveira R. The role of exopolymers in the attachment of Sphingomonas paucimobilis. *Biofouling* 2000; 16: 59-67.

[219] Jenkinson HF, Lappin-Scott HM. Biofilms adhere to stay. *Trends in Microbiology* 2001; 9: 9-10.

[220] Costerton JW. Overview of microbial biofilms. *Journal of Industrial Microbiology* 1995; 15: 137-140.

[221] Kokare CR, Chakraborty S, Khopade AN et al. Biofilm: Importance and applications. *Indian Journal of Biotechnology* 2009; 8: 159-168.

[222] Hwang G, Kang S, El-Din MG et al. Impact of an extracellular polymeric substance (EPS) precoating on the initial adhesion of Burkholderia cepacia and Pseudomonas aeruginosa. *Biofouling* 2012; 28: 525-538.

[223] Tsuneda S, Aikawa H, Hayashi H et al. Extracellular polymeric substances responsible for bacterial adhesion onto solid surface. *FEMS Microbiology Letters* 2003; 223: 287-292.

[224] Simões M, Bennett RN, Rosa EAS. Understanding antimicrobial activities of phytochemicals against multidrug resistant bacteria and biofilms. *Natural Product Reports* 2009; 26: 746-757.

[225] Sutherland IW. *Biofilm exopolysaccharides: A strong and sticky framework. Microbiology* 2001; 147: 3-9.

[226] Wingender J, Neu TR, Flemming HC. What are bacterial extracellular polymeric substances? *Microbial Extracellular Polymeric Substances* 1999: 1-19.

[227] Neu TR. Significance of bacterial surface-active compounds in interaction of bacteria with interfaces. *Microbiological Reviews* 1996; 60: 151-166.

[228] Whitchurch CB, Tolker-Nielsen T, Ragas PC et al. Extracellular DNA required for bacterial biofilm formation. *Science* 2002; 295: 1487.

[229] Klapper I, Rupp CJ, Cargo R et al. Viscoelastic fluid description of bacterial biofilm material properties. *Biotechnology and Bioengineering* 2002; 80: 289-296.

[230] Flemming HC, Wingender J, Mayer C et al. Cohesiveness in biofilm matrix polymers. *SGM Symposium Series* 59 2000: 87-105.

[231] Hussain M, Wilcox MH, White PJ. The slime of coagulase-negative staphylococci: Biochemistry and relation to adherence. *FEMS Microbiology Letters* 1993; 104: 191-207.

[232] Molobela IP, Cloete TE, Beukes M. Protease and amylase enzymes for biofilm removal and degradation of extracellular polymeric substances (EPS) produced by pseudomonas fluorescens bacteria. *African Journal of Microbiology Research* 2010; 4: 1515-1524.

[233] Leriche V, Sibille P, Carpentier B. Use of an enzyme-linked lectinsorbent assay to monitor the shift in polysaccharide composition in bacterial biofilms. *Applied and Environmental Microbiology* 2000; 66: 1851-1856.

[234] Donlan RM. Role of biofilms antimicrobial resistance. *ASAIO Journal* 2000; 46: S47-S52.

[235] Chmielewski RAN, Frank JF. Biofilm formation and control in food processing facilities. *Comprehensive Reviews in Food Science and Food Safety* 2003; 2: 22-32.

[236] McLean RJC, Whiteley M, Stickler DJ et al. Evidence of autoinducer activity in naturally occurring biofilms. *FEMS Microbiology Letters* 1997; 154: 259-263.

[237] O'Toole GA, Kolter R. Flagellar and twitching motility are necessary for Pseudomonas aeruginosa biofilm development. *Molecular Microbiology* 1998; 30: 295-304.

[238] Skerker JM, Berg HC. Direct observation of extension and retraction of type IV pili. *Proceedings of the National Academy of Sciences of the United States of America* 2001; 98: 6901-6904.

[239] Costerton JW, Ellis B, Lam K et al. Mechanism of electrical enhancement of efficacy of antibiotics in killing biofilm bacteria. *Antimicrobial Agents and Chemotherapy* 1994; 38: 2803-2809.

[240] Klausen M, Heydorn A, Ragas P et al. Biofilm formation by Pseudomonas aeruginosa wild type, flagella and type IV pili mutants. *Molecular Microbiology* 2003; 48: 1511-1524.

[241] Davey ME, O'Toole GA. Microbial biofilms: From ecology to molecular genetics. *Microbiology and Molecular Biology Reviews* 2000; 64: 847-867.

[242] Hooshangi S, Bentley WE. From unicellular properties to multicellular behavior: bacteria quorum sensing circuitry and applications. *Current Opinion in Biotechnology* 2008; 19: 550-555.

[243] Rasko DA, Moreira CG, De RL et al. Targeting QseC signaling and virulence for antibiotic development. *Science* 2008; 321: 1078-1080.

[244] Gomes IB, Simões M, Simões LC. An overview on the reactors to study drinking water biofilms. *Water Research* 2014; 62: 62-87.

[245] Choi J, Islam MS, Tumpa F et al. Biological fixed film. *Water Environment Research* 2013; 85: 1060-1091.

[246] Furey PC, Liess A. Substratum-associated microbiota. *Water Environment Research* 2013; 85: 1786-1827.

[247] Van Loosdrecht MCM, Picioreanu C, Heijnen JJ. A more unifying hypothesis for biofilm structures. *FEMS Microbiology Ecology* 1997; 24: 181-183.

[248] Picioreanu C, Van Loosdrecht MCM, Heijnen JJ. Two-dimensional model of biofilm detachment caused by internal stress from liquid flow. *Biotechnology and Bioengineering 2001*; 72: 205-218.

[249] Fonseca MV, Swanson MS. Nutrient salvaging and metabolism by the intracellular pathogen Legionella pneumophila. *Frontiers in Cellular and Infection Microbiology* 2014; 5.

[250] Lindsay D, Brözel VS, Von Holy A. Biofilm-spore response in Bacillus cereus and Bacillus subtilis during nutrient limitation. *Journal of Food Protection* 2006; 69: 1168-1172.

[251] Prigent-Combaret C, Lejeune P. Monitoring gene expression in biofilms. *Methods in Enzymology* 1999; 310: 56-79.

[252] Beloin C, Roux A, Ghigo JM. Escherichia coli biofilms. *Current Topics in Microbiology and Immunology 2008*; 322: 249-289.

[253] Bhomkar P, Materi W, Semenchenko V et al. Transcriptional response of E. coli upon FimH-mediated fimbrial adhesion. *Gene Regulation and Systems Biology* 2010; 2010: 1-17.

[254] Zogaj X, Nimtz M, Rohde M et al. The multicellular morphotypes of Salmonella typhimurium and Escherichia coli produce cellulose as the

second component of the extracellular matrix. *Molecular Microbiology* 2001; 39: 1452-1463.

[255] Bokranz W, Wang X, Tschäpe H et al. Expression of cellulose and curli fimbriae by Escherichia coli isolated from the gastrointestinal tract. *Journal of Medical Microbiology* 2005; 54: 1171-1182.

[256] Kai-Larsen Y, Lüthje P, Chromek M et al. Uropathogenic Escherichia coli modulates immune responses and its curli fimbriae interact with the antimicrobial peptide LL-37. *PLoS pathogens* 2010; 6.

[257] Danese PN, Pratt LA, Dove SL et al. The outer membrane protein, Antigen 43, mediates cell-to-cell interactions within Escherichia coli biofilms. *Molecular Microbiology* 2000; 37: 424-432.

[258] Agladze K, Wang X, Romeo T. Spatial periodicity of Eschenchia coli K-12 biofilm microstructure initiates during a reversible, polar attachment phase of development and requires the polysaccharide adhesin PGA. *Journal of Bacteriology* 2005; 187: 8237-8246.

[259] Cerca N, Maira-Litrán T, Jefferson KK et al. Protection against Escherichia coli infection by antibody to the Staphylococcus aureus poly-N-acetylglucosamine surface polysaccharide. *Proceedings of the National Academy of Sciences of the United States of America* 2007; 104: 7528-7533.

[260] Matsukawa M, Greenberg EP. Putative exopolysaccharide synthesis genes influence Pseudomonas aeruginosa biofilm development. *Journal of Bacteriology* 2004; 186: 4449-4456.

[261] Ma L, Jackson KD, Landry RM et al. Analysis of Pseudomonas aeruginosa conditional psl variants reveals roles for the psl polysaccharide in adhesion and maintaining biofilm structure postattachment. *Journal of Bacteriology* 2006; 188: 8213-8221.

[262] Harmsen M, Yang L, Pamp SJ et al. An update on Pseudomonas aeruginosa biofilm formation, tolerance, and dispersal. *FEMS Immunology and Medical Microbiology* 2010; 59: 253-268.

[263] Yang L, Barken KB, Skindersoe ME et al. Effects of iron on DNA release and biofilm development by Pseudomonas aeruginosa. *Microbiology* 2007; 153: 1318-1328.

[264] Allesen-Holm M, Barken KB, Yang L et al. A characterization of DNA release in Pseudomonas aeruginosa cultures and biofilms. *Molecular Microbiology* 2006; 59: 1114-1128.

[265] Barken KB, Pamp SJ, Yang L et al. Roles of type IV pili, flagellum-mediated motility and extracellular DNA in the formation of mature

multicellular structures in Pseudomonas aeruginosa biofilms. *Environmental Microbiology* 2008; 10: 2331-2343.

[266] Thomas VC, Hiromasa Y, Harms N et al. A fratricidal mechanism is responsible for eDNA release and contributes to biofilm development of Enterococcus faecalis. *Molecular Microbiology* 2009; 72: 1022-1036.

[267] Thomas VC, Thurlow LR, Boyle D et al. Regulation of autolysis-dependent extracellular DNA release by Enterococcus faecalis extracellular proteases influences biofilm development. *Journal of Bacteriology* 2008; 190: 5690-5698.

[268] Mohamed JA, Huang W, Nallapareddy SR et al. Influence of origin of isolates, especially endocarditis isolates, and various genes on biofilm formation by Enterococcus faecalis. *Infection and Immunity* 2004; 72: 3658-3663.

[269] Guiton PS, Hung CS, Kline KA et al. Contribution of autolysin and sortase A during Enterococcus faecalis DNA-dependent biofilm development. *Infection and Immunity* 2009; 77: 3626-3638.

[270] Sauer K, Camper AK, Ehrlich GD et al. Pseudomonas aeruginosa displays multiple phenotypes during development as a biofilm. *Journal of Bacteriology* 2002; 184: 1140-1154.

[271] Stoodley P, Cargo R, Rupp CJ et al. Biofilm material properties as related to shear-induced deformation and detachment phenomena. *Journal of Industrial Microbiology and Biotechnology* 2002; 29: 361-367.

[272] Kaplan JB, Ragunath C, Ramasubbu N et al. Detachment of Actinobacillus actinomycetemcomitans biofilm cells by an endogenous β-hexosaminidase activity. *Journal of Bacteriology 2003*; 185: 4693-4698.

[273] Kaplan JB, Ragunath C, Velliyagounder K et al. Enzymatic detachment of Staphylococcus epidermidis biofilms. *Antimicrobial Agents and Chemotherapy* 2004; 48: 2633-2636.

[274] Solano C, Echeverz M, Lasa I. Biofilm dispersion and quorum sensing. *Current Opinion in Microbiology* 2014; 18: 96-104.

[275] Wijman JGE, De Leeuw PPLA, Moezelaar R et al. Air-liquid interface biofilms of Bacillus cereus: Formation, sporulation, and dispersion. *Applied and Environmental Microbiology* 2007; 73: 1481-1488.

[276] Sauer K, Cullen MC, Rickard AH et al. Characterization of nutrient-induced dispersion in Pseudomonas aeruginosa PAO1 biofilm. *Journal of Bacteriology* 2004; 186: 7312-7326.

[277] Wilson S, Hamilton MA, Hamilton GC et al. Statistical quantification of detachment rates and size distributions of cell clumps from wild-type (PAO1) and cell signaling mutant (JP1) Pseudomonas aeruginosa biofilms. *Applied and Environmental Microbiology* 2004; 70: 5847-5852.

[278] Kaplan JB. Biofilm dispersal: mechanisms, clinical implications, and potential therapeutic uses. *Journal of dental research* 2010; 89: 205-218.

[279] Qiqi Y, Qiang H, Ibrahim HT. Review on moving bed biofilm processes. *Pakistan Journal of Nutrition* 2012; 11: 706-713.

[280] De Bivar Xavier J, Picioreanu C, Van Loosdrecht MCM. A general description of detachment for multidimensional modelling of biofilms. *Biotechnology and Bioengineering* 2005; 91: 651-669.

[281] Rochex A, Massé A, Escudié R et al. Influence of abrasion on biofilm detachment: Evidence for stratification of the biofilm. *Journal of Industrial Microbiology and Biotechnology* 2009; 36: 467-470.

[282] Morgenroth E, Wilderer PA. Influence of detachment mechanisms on competition in biofilms. *Water Research* 2000; 34: 417-426.

[283] So M, Naka D, Goel R et al. Modelling clogging and biofilm detachment in sponge carrier media. *Water Science and Technology* 2014; 69: 1298-1303.

[284] Sun S, Kjelleberg S, McDougald D. Relative Contributions of Vibrio Polysaccharide and Quorum Sensing to the Resistance of Vibrio cholerae to Predation by Heterotrophic Protists. *PLoS ONE* 2013; 8.

[285] Crutsinger GM. Bacterial biofilms in a 'genes-to-ecosystems' context. *Molecular Ecology* 2012; 21: 1545-1547.

[286] Laspidou CS. *Erosion probability for biofilm modeling: analysis of trends.* Desalination and Water Treatment 2013;.

[287] Bester E, Wolfaardt GM, Aznaveh NB et al. Biofilms' role in planktonic cell proliferation. *International Journal of Molecular Sciences* 2013; 14: 21965-21982.

[288] Ymele-Leki P, Ross JM. Erosion from Staphylococcus aureus biofilms grown under physiologically relevant fluid shear forces yields bacterial cells with reduced avidity to collagen. *Applied and Environmental Microbiology* 2007; 73: 1834-1841.

[289] Otto M. Staphylococcal biofilms. In, *Current Topics in Microbiology and Immunology;* 2008:207-228.

[290] Walter M, Safari A, Ivankovic A et al. Detachment characteristics of a mixed culture biofilm using particle size analysis. *Chemical Engineering Journal* 2013; 228: 1140-1147.

[291] Kumpel E, Nelson KL. Mechanisms affecting water quality in an intermittent piped water supply. *Environmental Science and Technology* 2014; 48: 2766-2775.

[292] Wang A, Lin B, Sleep BE et al. The Impact of Biofilm Growth on Transport of Escherichia coli O157:H7 in Sand. *Ground Water* 2011; 49: 20-31.

[293] Rupp CJ, Fux CA, Stoodley P. Viscoelasticity of Staphylococcus aureus biofilms in response to fluid shear allows resistance to detachment and facilitates rolling migration. *Applied and Environmental Microbiology* 2005; 71: 2175-2178.

[294] Anderson BN, Ding AM, Nilsson LM et al. Weak rolling adhesion enhances bacterial surface colonization. *Journal of Bacteriology* 2007; 189: 1794-1802.

[295] Fang H, Shang Q, Chen M et al. Changes in the critical erosion velocity for sediment colonized by biofilm. *Sedimentology* 2014; 61: 648-659.

[296] Jain A, Gupta Y, Agrawal R et al. Biofilms - A microbial life perspective: A critical review. *Critical Reviews in Therapeutic Drug Carrier Systems* 2007; 24: 393-443.

[297] Mathieu L, Bertrand I, Abe Y et al. Drinking water biofilm cohesiveness changes under chlorination or hydrodynamic stress. *Water Research* 2014; 55: 175-184.

[298] Dargahi M, Hosseinidoust Z, Tufenkji N et al. Investigating electrochemical removal of bacterial biofilms from stainless steel substrates. *Colloids and Surfaces B: Biointerfaces* 2014; 117: 152-157.

[299] Wang Z, Shen Y, Haapasalo M. Dental materials with antibiofilm properties. *Dental Materials* 2014; 30: e1-e16.

[300] Rmaile A, Carugo D, Capretto L et al. Removal of interproximal dental biofilms by high-velocity water microdrops. *Journal of Dental Research* 2014; 93: 68-73.

[301] De Souza EL, Meira QGS, Barbosa IM et al. Biofilm formation by Staphylococcus aureus from food contact surfaces in a meat-based broth and sensitivity to sanitizers. *Brazilian Journal of Microbiology* 2014; 45: 67-75.

[302] Kim J, Park HD, Chung S. Microfluidic approaches to bacterial biofilm formation. *Molecules* 2012; 17: 9818-9834.

[303] Hall-Stoodley L, Stoodley P. Biofilm formation and dispersal and the transmission of human pathogens. *Trends in Microbiology* 2005; 13: 7-10.

[304] Tolker-Nielsen T, Molin S. Spatial organization of microbial biofilm communities. *Microbial Ecology* 2000; 40: 75-84.

[305] James GA, Beaudette L, Costerton JW. Interspecies bacterial interactions in biofilms. *Journal of Industrial Microbiology* 1995; 15: 257-262.

[306] Lazăr V, Chifiriuc MC. Architecture and physiology of microbial biofilms. *Roumanian archives of microbiology and immunology* 2010; 69: 95-107.

[307] Durack DT. Experimental bacterial endocarditis. IV. Structure and evolution of very early lesions. *Journal of Pathology* 1975; 115: 81-89.

[308] Tunney MM, Jones DS, Gorman SP. Biofilm and biofilm-related encrustation of urinary tract devices. *Methods in Enzymology* 1999; 310: 558-566.

[309] Ascher DP, Shoupe BA, Maybee D et al. Persistent catheter-related bacteremia: Clearance with antibiotics and urokinase. *Journal of Pediatric Surgery* 1993; 28: 627-629.

[310] Bouza E, Burillo A, Muñoz P. Catheter-related infections: Diagnosis and intravascular treatment. *Clinical Microbiology and Infection* 2002; 8: 265-274.

[311] Hamadi F, Asserne F, Elabed S et al. Adhesion of Staphylococcus aureus on stainless steel treated with three types of milk. *Food Control* 2014; 38: 104-108.

[312] Mureddu A, Mazza R, Fois F et al. Listeria monocytogenes persistence in ready-to-eat sausages and in processing plants. *Italian Journal of Food Safety* 2014; 3: 12-15.

[313] Daniels R, Vanderleyden J, Michiels J. Quorum sensing and swarming migration in bacteria. *FEMS Microbiology Reviews* 2004; 28: 261-289.

[314] Davies DG, Marques CNH. A fatty acid messenger is responsible for inducing dispersion in microbial biofilms. *Journal of Bacteriology* 2009; 191: 1393-1403.

[315] Garg N, Manchanda G, Kumar A. Bacterial quorum sensing: Circuits and applications. Antonie van Leeuwenhoek, *International Journal of General and Molecular Microbiology* 2014; 105: 289-305.

[316] Davies DG, Parsek MR, Pearson JP et al. The involvement of cell-to-cell signals in the development of a bacterial biofilm. *Science* 1998; 280: 295-298.

[317] Romero D. Bacterial determinants of the social behavior of Bacillus subtilis. *Research in Microbiology* 2013; 164: 788-798.

[318] Jiang T, Li M. Quorum sensing inhibitors: A patent review. *Expert Opinion on Therapeutic Patents* 2013; 23: 867-894.

[319] Bai AJ, Rai VR. Bacterial Quorum Sensing and Food Industry. *Comprehensive Reviews in Food Science and Food Safety* 2011; 10: 183-193.

[320] Smith JL, Fratamico PM, Novak JS. Quorum sensing: A primer for food microbiologists. *Journal of Food Protection* 2004; 67: 1053-1070.

[321] Fuqua C, Winans SC, Greenberg EP. Census and consensus in bacterial ecosystems: The LuxR-LuxI family of quorum-sensing transcriptional regulators. *Annual Review of Microbiology* 1996; 50: 727-751.

[322] West SA, Winzer K, Gardner A et al. Quorum sensing and the confusion about diffusion. *Trends in Microbiology* 2012; 20: 586-594.

[323] Fuqua WC, Winans SC, Greenberg EP. Quorum sensing in bacteria: The LuxR-LuxI family of cell density- responsive transcriptional regulators. *Journal of Bacteriology* 1994; 176: 269-275.

[324] Fuqua C, Greenberg EP. Listening in on bacteria: Acyl-homoserine lactone signalling. *Nature Reviews Molecular Cell Biology* 2002; 3: 685-695.

[325] Bandara HMHN, Lam OLT, Jin LJ et al. Microbial chemical signaling: A current perspective. *Critical Reviews in Microbiology* 2012; 38: 217-249.

[326] Biradar B, Devi P. Quorum sensing in plaque biofilms: Challenges and future prospects. *Journal of Contemporary Dental Practice* 2011; 12: 479-485.

[327] Eberhard A, Burlingame AL, Eberhard C et al. Structural identification of autoinducer of Photobacterium fischeri luciferase. *Biochemistry* 1981; 20: 2444-2449.

[328] Toyofuku M, Uchiyama H, Nomura N. Social behaviours under anaerobic conditions in Pseudomonas aeruginosa. *International Journal of Microbiology 2012;.*

[329] Cappitelli F, Polo A, Villa F. Biofilm Formation in Food Processing Environments is Still Poorly Understood and Controlled. *Food Engineering Reviews* 2014; 6: 29-42.

[330] Da Silva EP, De Martinis ECP. Current knowledge and perspectives on biofilm formation: The case of Listeria monocytogenes. *Applied Microbiology and Biotechnology* 2013; 97: 957-968.

[331] Jakobsen TH, Bragason SK, Phipps RK et al. Food as a source for quorum sensing inhibitors: Iberin from horseradish revealed as a quorum sensing inhibitor of Pseudomonas aeruginosa. *Applied and Environmental Microbiology* 2012; 78: 2410-2421.

[332] Galloway WRJD, Hodgkinson JT, Bowden S et al. Applications of small molecule activators and inhibitors of quorum sensing in Gram-negative bacteria. *Trends in Microbiology* 2012; 20: 449-458.

[333] Rasch M, Andersen JB, Nielsen KF et al. Involvement of bacterial quorum-sensing signals in spoilage of bean sprouts. *Applied and Environmental Microbiology* 2005; 71: 3321-3330.

[334] Silagyi K, Kim SH, Martin Lo Y et al. Production of biofilm and quorum sensing by Escherichia coli O157:H7 and its transfer from contact surfaces to meat, poultry, ready-to-eat deli, and produce products. *Food Microbiology* 2009; 26: 514-519.

[335] Moons P, Van Houdt R, Aertsen A et al. Role of quorum sensing and antimicrobial component production by Serratia plymuthica in formation of biofilms, including mixed biofilms with Escherichia coli. *Applied and Environmental Microbiology* 2006; 72: 7294-7300.

[336] Lu L, Hume ME, Pillai SD. Autoinducer-2-like activity on vegetable produce and its potential involvement in bacterial biofilm formation on tomatoes. *Foodborne Pathogens and Disease* 2005; 2: 242-249.

[337] Gram L, Christensen AB, Ravn L et al. Production of acylated homoserine lactones by psychrotrophic members of the Enterobacteriaceae isolated from foods. *Applied and Environmental* Microbiology 1999; 65: 3458-3463.

[338] Van Houdt R, Moons P, Jansen A et al. Genotypic and phenotypic characterization of a biofilm-forming Serratia plymuthica isolate from a raw vegetable processing line. *FEMS Microbiology Letters* 2005; 246: 265-272.

[339] Ehlers LJ, Bouwer EJ. RP4 plasmid transfer among species of Pseudomonas in a biofilm reactor. *Water Science and Technology* 1999; 39: 163-171.

[340] Roberts AP, Pratten J, Wilson M et al. Transfer of a conjugative transposon, Tn5397 in a model oral biofilm. *FEMS Microbiology Letters* 1999; 177: 63-66.

[341] Hausner M, Wuertz S. High rates of conjugation in bacterial biofilms as determined by quantitative in situ analysis. *Applied and Environmental Microbiology* 1999; 65: 3710-3713.

[342] Aminov RI. Horizontal gene exchange in environmental microbiota. *Frontiers in Microbiology* 2011; 2.

[343] Montanaro L, Poggi A, Visai L et al. Extracellular DNA in biofilms. *International Journal of Artificial Organs* 2011; 34: 824-831.

[344] Lamont RJ, Bryers JD. Biofilm-induced gene expression and gene transfer. In, *Methods in Enzymology*; 2001:84-94.

[345] Ghigo JM. Natural conjugative plasmids induce bacterial biofilm development. *Nature* 2001; 412: 442-445.

[346] Verraes C, Van Boxstael S, Van Meervenne E et al. Antimicrobial resistance in the food chain: A review. *International Journal of Environmental Research and Public Health* 2013; 10: 2643-2669.

[347] Mitri S, Richard Foster K. The genotypic view of social interactions in microbial communities. In, *Annual Review of Genetics*; 2013:247-273.

[348] Murga R, Miller JM, Donlan RM. Biofilm formation by gram-negative bacteria on central venous catheter connectors: Effect of conditioning films in a laboratory model. *Journal of Clinical Microbiology* 2001; 39: 2294-2297.

[349] McLaughlin-Borlace L, Stapleton F, Matheson M et al. Bacterial biofilm on contact lenses and lens storage cases in wearers with microbial keratitis. *Journal of Applied Microbiology* 1998; 84: 827-838.

[350] Derlon N, Peter-Varbanets M, Scheidegger A et al. Predation influences the structure of biofilm developed on ultrafiltration membranes. *Water Research* 2012; 46: 3323-3333.

[351] Stewart PS, Camper AK, Handran SD et al. Spatial distribution and coexistence of Klebsiella pneumoniae and Pseudomonas aeruginosa in biofilms. *Microbial Ecology* 1997; 33: 2-10.

[352] Friman VP, Diggle SP, Buckling A. Protist predation can favour cooperation within bacterial species. *Biology letters* 2013; 9: 20130548.

[353] Raad II, Sabbagh MF, Rand KH et al. Quantitative tip culture methods and the diagnosis of central venous catheter-related infections. *Diagnostic Microbiology and Infectious Disease* 1992; 15: 13-20.

[354] Wirtanen G, Husmark U, Mattila-Sandholm T. Microbial evaluation of the biotransfer potential from surfaces with Bacillus biofilms after rinsing and cleaning procedures in closed food- processing systems. *Journal of Food Protection* 1996; 59: 727-733.

[355] Camper AK, Warnecke M, Jones WL et al. *Pathogens in Model Distribution System Biofilms* 1998;.

[356] Stark RM, Gerwig GJ, Pitman RS et al. Biofilm formation by Helicobacter pylori. *Letters in Applied Microbiology* 1999; 28: 121-126.

[357] Wintermute EH, Silver PA. Dynamics in the mixed microbial concourse. *Genes and Development* 2010; 24: 2603-2614.

[358] Morris AR, Visick KL. Control of biofilm formation and colonization in Vibrio fischeri: A role for partner switching? *Environmental Microbiology* 2010; 12: 2051-2059.

[359] Falsetta ML, Klein MI, Colonne PM et al. Symbiotic relationship between Streptococcus mutans and Candida albicans synergizes virulence of plaque biofilms in vivo. *Infection and Immunity* 2014; 82: 1968-1981.

[360] Gjermansen M, Ragas P, Sternberg C et al. Characterization of starvation-induced dispersion in Pseudomonas putida biofilms. *Environmental Microbiology* 2005; 7: 894-906.

[361] Jackson KD, Starkey M, Kremer S et al. Identification of psl, a locus encoding a potential exopolysaccharide that is essential for Pseudomonas aeruginosa PAO1 biofilm formation. *Journal of Bacteriology* 2004; 186: 4466-4475.

[362] Sauer FG, Remaut H, Hultgren SJ et al. Fiber assembly by the chaperone-usher pathway. *Biochimica et Biophysica Acta - Molecular Cell Research* 2004; 1694: 259-267.

[363] Thormann KM, Saville RM, Shukla S et al. Induction of rapid detachment in Shewanella oneidensis MR-1 biofilms. *Journal of Bacteriology* 2005; 187: 1014-1021.

[364] Wang Y, Zheng SJ, Pei LY et al. Nutrient release, recovery and removal from waste sludge of a biological nutrient removal system. *Environmental Technology* (United Kingdom) 2014;.

[365] Donlan RM, Pipes WO, Yohe TL. Biofilm formation on cast iron substrata in water distribution systems. *Water Research* 1994; 28: 1497-1503.

[366] Fera P, Siebel MA, Characklis WG et al. Seasonal variations in bacterial colonization of stainless steel, aluminium and polycarbonate surfaces in a seawater flow system. *Biofouling* 1989; 1: 251-261.

[367] Orvain F, De Crignis M, Guizien K et al. Tidal and seasonal effects on the short-term temporal patterns of bacteria, microphytobenthos and exopolymers in natural intertidal biofilms (Brouage, France). *Journal of Sea Research* 2014;.

[368] Dunne Jr WM. Bacterial adhesion: Seen any good biofilms lately? *Clinical Microbiology Reviews* 2002; 15: 155-166.

[369] Murga R, Forster TS, Brown E et al. Role of biofilms in the survival of Legionella pneumophila in a model potable-water system. *Microbiology* 2001; 147: 3121-3126.
[370] Cowan MM, Warren TM, Fletcher M. Mixed species colonization of solid surfaces in laboratory biofilms. *Biofouling* 1991; 3: 23-34.
[371] Stoodley P, Dodds I, Boyle JD et al. Influence of hydrodynamics and nutrients on biofilm structure. *Journal of Applied Microbiology Symposium Supplement* 1999; 85: 19S-28S.
[372] Kim KY, Frank JF. Effect of growth nutrients on attachment of Listeria monocytogenes to stainless steel. *J Food Prot* 1994; 57: 720-726.
[373] Kelly PT, He Z. Nutrients removal and recovery in bioelectrochemical systems: A review. *Bioresource Technology* 2014; 153: 351-360.
[374] Greenman J, Saad S, Hewett K et al. Review: In vitro biofilm models for studying oral malodour. *Flavour and Fragrance Journal* 2013; 28: 212-222.
[375] Abbas HA, Serry FM, El-Masry EM. Biofilms: The microbial castle of resistance. *Research Journal of Pharmacy and Technology* 2013; 6: 1-3.
[376] Shao C, Sun Y, Wang N et al. Changes of proteome components of Helicobacter pylori biofilms induced by serum starvation. *Molecular Medicine Reports* 2013; 8: 1761-1766.
[377] Moltz AG, Martin SE. Formation of biofilms by Listeria monocytogenes under various growth conditions. *Journal of Food Protection* 2005; 68: 92-97.
[378] Stepanović S, Ćirković I, Ranin L et al. Biofilm formation by Salmonella spp. and Listeria monocytogenes on plastic surface. *Letters in Applied Microbiology* 2004; 38: 428-432.
[379] Rochex A, Godon JJ, Bernet N et al. Role of shear stress on composition, diversity and dynamics of biofilm bacterial communities. *Water Research* 2008; 42: 4915-4922.
[380] Costerton JW, Lappin-Scott HM. Introduction to microbial biofilms. *Microbial Biofilms* 1995: 1-11.
[381] Lagha R, Bellon-Fontaine MN, Renault M et al. Impact of long-term starvation on adhesion to and biofilm formation on stainless steel 316 L and gold surfaces of Salmonella enterica serovar Typhimurium. *Annals of Microbiology* 2014;.
[382] Ning Y, Hu X, Ling J et al. Candida albicans survival and biofilm formation under starvation conditions. *International Endodontic Journal* 2013; 46: 62-70.

[383] Marshall KC. The importance of studying microbial cell surfaces. *Microbial Cell Surface Analysis* 1991: 3-19.

[384] Wong ACL. Influence of culture conditions on biofilm formation by Escherichia coli O157:H7. *International Journal of Food Microbiology* 1995; 26: 147-164.

[385] Amato SM, Brynildsen MP. Nutrient transitions are a source of persisters in Escherichia coli biofilms. *PLoS ONE* 2014; 9.

[386] Stewart PS. Mini-review: Convection around biofilms. *Biofouling* 2012; 28: 187-198.

[387] Vigolo D, Al-Housseiny TT, Shen Y et al. Flow dependent performance of microfluidic microbial fuel cells. *Physical Chemistry Chemical Physics* 2014; 16: 12535-12543.

[388] Wang H, Sodagari M, Ju LK et al. Effects of shear on initial bacterial attachment in slow flowing systems. *Colloids and Surfaces B: Biointerfaces* 2013; 109: 32-39.

[389] Hall-Stoodley L, Costerton JW, Stoodley P. Bacterial biofilms: From the natural environment to infectious diseases. *Nature Reviews Microbiology 2004*; 2: 95-108.

[390] Isberg RR, Barnes P. Dancing with the host: Flow-dependent bacterial adhesion. *Cell* 2002; 110: 1-4.

[391] Moreira JMR, Teodósio JS, Silva FC et al. Influence of flow rate variation on the development of Escherichia coli biofilms. *Bioprocess and Biosystems Engineering* 2013; 36: 1787-1796.

[392] Salta M, Capretto L, Carugo D et al. Life under flow: A novel microfluidic device for the assessment of anti-biofilm technologies. *Biomicrofluidics* 2013; 7.

[393] Boles BR, Horswill AR. agr-mediated dispersal of Staphylococcus aureus biofilms. *PLoS Pathogens* 2008; 4.

[394] Davey ME, Caiazza NC, O'Toole GA. Rhamnolipid surfactant production affects biofilm architecture in Pseudomonas aeruginosa PAO1. *Journal of Bacteriology* 2003; 185: 1027-1036.

[395] Purevdorj B, Costerton JW, Stoodley P. Influence of hydrodynamics and cell signaling on the structure and behavior of Pseudomonas aeruginosa biofilms. *Applied and Environmental Microbiology* 2002; 68: 4457-4464.

[396] Lecuyer S, Rusconi R, Shen Y et al. Shear stress increases the residence time of adhesion of Pseudomonas aeruginosa. *Biophysical Journal* 2011; 100: 341-350.

[397] Simões M, Pereira MO, Sillankorva S et al. The effect of hydrodynamic conditions on the phenotype of Pseudomonas fluorescens biofilms. *Biofouling* 2007; 23: 249-258.

[398] Perni S, Jordan SJ, Andrew PW et al. Biofilm development by Listeria innocua in turbulent flow regimes. *Food Control* 2006; 17: 875-883.

[399] Mehdi Salek M, Jones SM, Martinuzzi RJ. The influence of flow cell geometry related shear stresses on the distribution, structure and susceptibility of Pseudomonas aeruginosa 01 biofilms. *Biofouling* 2009; 25: 711-725.

[400] Lee JH, Kaplan JB, Lee WY. Microfluidic devices for studying growth and detachment of Staphylococcus epidermidis biofilms. *Biomedical Microdevices* 2008; 10: 489-498.

[401] Tsai YP. Impact of flow velocity on the dynamic behaviour of biofilm bacteria. *Biofouling* 2005; 21: 267-277.

[402] Dourou D, Beauchamp CS, Yoon Y et al. Attachment and biofilm formation by Escherichia coli O157:H7 at different temperatures, on various food-contact surfaces encountered in beef processing. *International Journal of Food Microbiology* 2011; 149: 262-268.

[403] Arcangeli JP, Arvin E. Growth of an aerobic and an anoxic toluene-degrading biofilm - A comparative study. In, *Water Science and Technology;* 1995:125-132.

[404] Villaverde S, Mirpuri R, Lewandowski Z et al. Study of toluene degradation kinetics in a flat plate vapor phase bioreactor using oxygen microsensors. In, *Water Science and Technology;* 1997:77-84.

[405] Santegoeds CM, Ferdelman TG, Muyzer G et al. Structural and functional dynamics of sulfate-reducing populations in bacterial biofilms. *Applied and Environmental Microbiology* 1998; 64: 3731-3739.

[406] Kroukamp O, Wolfaardt GM. CO2 production as an indicator of biofilm metabolism. *Applied and Environmental Microbiology* 2009; 75: 4391-4397.

[407] Skolimowski M, Nielsen MW, Emnéus J et al. Microfluidic dissolved oxygen gradient generator biochip as a useful tool in bacterial biofilm studies. *Lab on a Chip - Miniaturisation for Chemistry and Biology* 2010; 10: 2162-2169.

[408] Applegate LA, Luscher P, Tyrrell RM. Induction of heme oxygenase: A general response to oxidant stress in cultured mammalian cells. *Cancer Research* 1991; 51: 974-978.

[409] Applegate DH, Bryers JD. Effects of carbon and oxygen limitations and calcium concentrations on biofilm removal processes. *Biotechnology and Bioengineering* 1991; 37: 17-25.
[410] Mun S, Jeong JS, Kim J et al. Inactivation of Pseudomonas aeruginosa biofilm by dense phase carbon dioxide. *Biofouling* 2009; 25: 473-479.
[411] Rothenbacher FP, Zhu J. Efficient responses to host and bacterial signals during Vibrio cholerae colonization. *Gut Microbes* 2014; 5.
[412] Pöllänen MT, Paino A, Ihalin R. Environmental stimuli shape biofilm formation and the virulence of periodontal pathogens. *International Journal of Molecular Sciences* 2013; 14: 17221-17237.
[413] Goller CC, Romeo T. Environmental influences on biofilm development. In, *Current Topics in Microbiology and Immunology;* 2008:37-66.
[414] Wirthlin MR, Marshall Jr GW, Rowland RW. Formation and decontamination of biofilms in dental unit waterlines. *Journal of Periodontology* 2003; 74: 1595-1609.
[415] Booth RB, Mead LC. *US Patent.* 1969: 730.
[416] Olsen A, Jonsson A, Normark S. Fibronectin binding mediated by a novel class of surface organelles on Escherichia coli. *Nature* 1989; 338: 652-655.
[417] Serefko AD, Poleszak EJ, Malm A. Candida albicans denture biofilm and its clinical significance. *Polish Journal of Microbiology* 2012; 61: 161-167.
[418] Plette ACC, van Riemsdijk WH, Benedetti MF et al. pH Dependent Charging Behavior of Isolated Cell Walls of a Gram-Positive Soil Bacterium. *Journal of Colloid And Interface Science* 1995; 173: 354-363.
[419] Oliveira DR. Physico-chemical aspects of adhesion. *Biofilms Science and Technology* 1992: 45-58.
[420] Stepanović S, Ćirković I, Mijač V et al. Influence of the incubation temperature, atmosphere and dynamic conditions on biofilm formation by Salmonella spp. *Food Microbiology* 2003; 20: 339-343.
[421] Villanueva VD, Font J, Schwartz T et al. Biofilm formation at warming temperature: Acceleration of microbial colonization and microbial interactive effects. *Biofouling* 2011; 27: 59-71.
[422] Burgess SA, Lindsay D, Flint SH. Thermophilic bacilli and their importance in dairy processing. *International Journal of Food Microbiology* 2010; 144: 215-225.

[423] Fletcher M, Pringle JH. The effect of surface free energy and medium surface tension on bacterial attachment to solid surfaces. *Journal of Colloid And Interface Science* 1985; 104: 5-14.
[424] Herald PJ, Zottola EA. Attachment of Listeria monocytogenes to stainless steel surfaces at various temperatures and pH values. *J Food Sci* 1988; 53: 1549-1552.
[425] Marion-Ferey K, Pasmore M, Stoodley P et al. Biofilm removal from silicone tubing: An assessment of the efficacy of dialysis machine decontamination procedures using an in vitro model. *Journal of Hospital Infection* 2003; 53: 64-71.
[426] Christersson CE, Glantz PO, Baier RE. Role of temperature and shear forces on microbial detachment. *Scandinavian Journal of Dental Research* 1988; 96: 91-98.
[427] Hood SK, Zottola EA. Isolation and identification of adherent gram-negative microorganisms from four meat-processing facilities. *Journal of Food Protection* 1997; 60: 1135-1138.
[428] Smoot LM, Pierson MD. Influence of environmental stress on the kinetics and strength of attachment of Listeria monocytogenes Scott A to Buna-N rubber and stainless steel. *Journal of Food Protection* 1998; 61: 1286-1292.
[429] Hassan AN, Birt DM, Frank JF. Behavior of Listeria monocytogenes in a Pseudomonas putida Biofilm on a Condensate-Forming Surface. *Journal of Food Protection* 2004; 67: 322-327.
[430] Fry JC. 2 Direct Methods and Biomass Estimation. *Methods in Microbiology* 1990; 22: 41-85.
[431] Beyenal H, Lewandowski Z, Harkin G. Quantifying biofilm structure: Facts and fiction. *Biofouling* 2004; 20: 1-23.
[432] Smirnova TA, Didenko LV, Azizbekyan RR et al. Structural and functional characteristics of bacterial biofilms. *Microbiology* 2010; 79: 413-423.
[433] Fletcher M. Attachment of Pseudomonas fluorescens to glass and influence of electrolytes on bacterium-substratum separation distance. *Journal of Bacteriology* 1988; 170: 2027-2030.
[434] Corcuera MT, Gómez-Lus ML, Gómez-Aguado F et al. Morphological plasticity of Streptococcus oralis isolates for biofilm production, invasiveness, and architectural patterns. *Archives of Oral Biology* 2013; 58: 1584-1593.
[435] Kennedy P, Brammah S, Wills E. Burns, biofilm and a new appraisal of burn wound sepsis. *Burns* 2010; 36: 49-56.

[436] Rühs PA, Böcker L, Inglis RF et al. Studying bacterial hydrophobicity and biofilm formation at liquid–liquid interfaces through interfacial rheology and pendant drop tensiometry. *Colloids and Surfaces B: Biointerfaces* 2014; 117: 174-184.

[437] Kodjikian L, Burillon C, Lina G et al. Biofilm formation on intraocular lenses by a clinical strain encoding the ica locus: A scanning electron microscopy study. *Investigative Ophthalmology and Visual Science* 2003; 44: 4382-4387.

[438] Ganderton L, Chawla J, Winters C et al. Scanning electron microscopy of bacterial biofilms on indwelling bladder catheters. *European Journal of Clinical Microbiology and Infectious Diseases* 1992; 11: 789-796.

[439] Raad I, Costerton W, Sabharwal U et al. Ultrastructural analysis of indwelling vascular catheters: A quantitative relationship between luminal colonization and duration of placement. *Journal of Infectious Diseases* 1993; 168: 400-407.

[440] Stickler DJ, Morris NS, McLean RJC et al. Biofilms on indwelling urethral catheters produce quorum sensing signal molecules in situ and in vitro. *Applied and Environmental Microbiology 1998*; 64: 3486-3490.

[441] Nickel JC, Costerton JW. Bacterial biofilms and catheters: A key to understanding bacterial strategies in catheter-associated urinary tract infections. *Can J Infect Dis* 1992; 3: 619-624.

[442] Ansari MJ, Al-Ghamdi A, Usmani S et al. Effect of Jujube Honey on Candida albicans Growth and Biofilm Formation. *Archives of Medical Research* 2013; 44: 352-360.

[443] Lou Z, Song X, Hong Y et al. Separation and enrichment of burdock leaf components and their inhibition activity on biofilm formation of E. coli. *Food Control* 2013; 32: 270-274.

[444] Brambilla E, Ionescu A, Mazzoni A et al. Hydrophilicity of dentin bonding systems influences in vitro Streptococcus mutans biofilm formation. *Dental Materials;*.

[445] Christensen BB, Sternberg C, Andersen JB et al. Molecular tools for study of biofilm physiology. *Methods in Enzymology* 1999; 310: 20-42.

[446] Suwarno SR, Chen X, Chong TH et al. Biofouling in reverse osmosis processes: The roles of flux, crossflow velocity and concentration polarization in biofilm development. *Journal of Membrane* Science 2014; 467: 116-125.

[447] Karygianni L, Follo M, Hellwig E et al. Microscope-based imaging platform for large-scale analysis of oral biofilms. *Applied and Environmental Microbiology* 2012; 78: 8703-8711.

[448] Oliveira A, Cunha MDLRS. *Comparison of methods for the detection of biofilm production in coagulase-negative staphylococci.* BMC Research Notes 2010; 3.

[449] Bridier A, Dubois-Brissonnet F, Boubetra A et al. The biofilm architecture of sixty opportunistic pathogens deciphered using a high throughput CLSM method. *Journal of Microbiological Methods* 2010; 82: 64-70.

[450] Danielsen KA, Eskeland Ø, Fridrich-Aas K et al. Bacterial biofilms in patients with chronic rhinosinusitis: A confocal scanning laser microscopy study. *Rhinology* 2014; 52: 150-155.

[451] Wu D, Fan W, Kishen A et al. Evaluation of the antibacterial efficacy of silver nanoparticles against Enterococcus faecalis biofilm. *Journal of Endodontics* 2014; 40: 285-290.

[452] Oh YJ, Jo W, Yang Y et al. Influence of culture conditions on Escherichia coli O157:H7 biofilm formation by atomic force microscopy. *Ultramicroscopy* 2007; 107: 869-874.

[453] Dorobantu LS, Goss GG, Burrell RE. Atomic force microscopy: A nanoscopic view of microbial cell surfaces. *Micron* 2012; 43: 1312-1322.

[454] Bremer PJ, Geesey GG, Drake B. Atomic force microscopy examination of the topography of a hydrated bacterial biofilm on a copper surface. *Current Microbiology* 1992; 24: 223-230.

[455] Oh YJ, Lee NR, Jo W et al. Effects of substrates on biofilm formation observed by atomic force microscopy. *Ultramicroscopy* 2009; 109: 874-880.

[456] Wright CJ, Powell LC, Johnson DJ et al. MICROSCOPY | Atomic Force Microscopy. In: Batt CA, Tortorello ML, Hrsg. *Encyclopedia of Food Microbiology* (Second Edition). Oxford: Academic Press; 2014:666-675.

[457] Bartke A, Wright JC, Mattison JA et al. Extending the lifespan of long-lived mice. *Nature* 2001; 414: 412.

[458] Longo G, Kasas S. Effects of antibacterial agents and drugs monitored by atomic force microscopy. *Wiley Interdisciplinary Reviews: Nanomedicine and Nanobiotechnology* 2014; 6: 230-244.

[459] Yang X, Yang W, Wang Q et al. Atomic force microscopy investigation of the characteristic effects of silver ions on Escherichia coli and Staphylococcus epidermidis. *Talanta* 2010; 81: 1508-1512.

[460] Tsoligkas AN, Bowen J, Winn M et al. Characterisation of spin coated engineered Escherichia coli biofilms using atomic force microscopy. *Colloids and Surfaces B: Biointerfaces* 2012; 89: 152-160.

[461] Kruszewski KM, Nistico L, Longwell MJ et al. Reducing Staphylococcus aureus biofilm formation on stainless steel 316L using functionalized self-assembled monolayers. *Materials Science and Engineering:* C 2013; 33: 2059-2069.

[462] Ribeiro MMB, Franquelim HG, Torcato IM et al. Antimicrobial properties of analgesic kyotorphin peptides unraveled through atomic force microscopy. *Biochemical and Biophysical Research Communications* 2012; 420: 676-679.

[463] Van Der Mei HC, Busscher HJ, Bos R et al. Direct probing by atomic force microscopy of the cell surface softness of a fibrillated and nonfibrillated oral streptococcal strain. *Biophysical Journal* 2000; 78: 2668-2674.

[464] Gorman SP, McGovern JG, David Woolfson A et al. The concomitant development of poly(vinyl chloride)-related biofilm and antimicrobial resistance in relation to ventilator-associated pneumonia. *Biomaterials* 2001; 22: 2741-2747.

[465] Jones DS, McGovern JG, Woolfson AD et al. Physicochemical characterization of hexetidine-impregnated endotracheal tube poly(vinyl chloride) and resistance to adherence of respiratory bacterial pathogens. *Pharmaceutical Research* 2002; 19: 818-824.

[466] Schaefer AL, Greenberg EP, Parsek MR. [4] Acylated homoserine lactone detection in pseudomonas aeruginosa Biofilms by radiolabel assay. In: Ron JD, Hrsg. *Methods in Enzymology:* Academic Press; 2001:41-47.

[467] Okabe S, Satoh H, Kindaichi T. Chapter Seven - A Polyphasic Approach to Study Ecophysiology of Complex Multispecies Nitrifying Biofilms. In: Martin GKaLYS, Hrsg. *Methods in Enzymology:* Academic Press; 2011:163-184.

[468] Chattopadhyay S, Saha Das S, Chandra S et al. Synthesis and evaluation of 99mTc-moxifloxacin, a potential infection specific imaging agent. *Applied Radiation and Isotopes* 2010; 68: 314-316.

[469] Ross MK, Pegram RA. [35S]-Labeling of the Salmonella typhimurium glutathione pool to assess glutathione-mediated DNA binding by 1,2-dibromoethane. *Chemico-Biological Interactions* 2003; 146: 39-49.

[470] Christensen GD, Simpson WA, Younger JJ et al. Adherence of coagulase-negative staphylococci to plastic tissue culture plates: A

quantitative model for the adherence of staphylococci to medical devices. *Journal of Clinical Microbiology* 1985; 22: 996-1006.

[471] Pantanella F, Valenti P, Natalizi T et al. Analytical techniques to study microbial biofilm on abiotic surfaces: pros and cons of the main techniques currently in use. Annali di igiene: *Medicina preventiva e di comunità* 2013; 25: 31-42.

[472] Li X, Yan Z, Xu J. Quantitative variation of biofilms among strains in natural populations of Candida albicans. *Microbiology* 2003; 149: 353-362.

[473] Maki DG, Weise CE, Sarafin HW. A semiquantitative culture method for identifying intravenous catheter related infection. *New England Journal of Medicine* 1977; 296: 1305-1309.

[474] Green JBD, Fulghum T, Nordhaus MA. Review of immobilized antimicrobial agents and methods for testing. *Biointerphases* 2011; 6: MR13-MR28.

[475] Morris CE, Monier JM, Jacques MA. A technique to quantify the population size and composition of the biofilm component in communities of bacteria in the phyllosphere. *Applied and Environmental Microbiology* 1998; 64: 4789-4795.

[476] Lindsay D, von Holy A. Bacterial biofilms within the clinical setting: what healthcare professionals should know. *Journal of Hospital Infection* 2006; 64: 313-325.

[477] Hassan A, Usman J, Kaleem F et al. Evaluation of different detection methods of biofilm formation in the clinical isolates. *Brazilian Journal of Infectious Diseases* 2011; 15: 305-311.

[478] Sharvari SA, Chitra PG. Evaluation of different detection methods of biofilm formation in clinical isolates of Staphylococci. *International Journal of Pharma and Bio Sciences* 2012; 3: 724-733.

[479] Khamisse E, Firmesse O, Christieans S et al. Impact of cleaning and disinfection on the non-culturable and culturable bacterial loads of food-contact surfaces at a beef processing plant. *International Journal of Food Microbiology* 2012; 158: 163-168.

[480] Dinu LD, Bach S. Induction of viable but nonculturable Escherichia coli O157:H7 in the phyllosphere of lettuce: A food safety risk factors. *Applied and Environmental Microbiology* 2011; 77: 8295-8302.

[481] Lindbäck T, Rottenberg ME, Roche SM et al. The ability to enter into an avirulent viable but non-culturable (VBNC) form is widespread among Listeria monocytogenes isolates from salmon, patients and environment. *Veterinary Research* 2010; 41.

[482] Pasquaroli S, Zandri G, Vignaroli C et al. Antibiotic pressure can induce the viable but non-culturable state in Staphylococcus aureus growing in biofilms. *Journal of Antimicrobial Chemotherapy* 2013; 68: 1812-1817.

[483] Senoh M, Ghosh-Banerjee J, Ramamurthy T et al. Conversion of viable but nonculturable enteric bacteria to culturable by co-culture with eukaryotic cells. *Microbiology and Immunology* 2012; 56: 342-345.

[484] Costerton JW, Post JC, Ehrlich GD et al. New methods for the detection of orthopedic and other biofilm infections. *FEMS Immunology and Medical Microbiology* 2011; 61: 133-140.

[485] Nguyen CT, Jung W, Kim J et al. Noninvasive in vivo optical detection of biofilm in the human middle ear. *Proceedings of the National Academy of Sciences of the United States of America* 2012; 109: 9529-9534.

[486] Lu X, Al-Qadiri HM, Lin M et al. Application of Mid-infrared and Raman Spectroscopy to the Study of Bacteria. *Food and Bioprocess Technology* 2011; 4: 919-935.

[487] Jun W, Kim MS, Cho BK et al. Microbial biofilm detection on food contact surfaces by macro-scale fluorescence imaging. *Journal of Food Engineering* 2010; 99: 314-322.

[488] Kim YW, Sardari SE, Meyer MT et al. An ALD aluminum oxide passivated Surface Acoustic Wave sensor for early biofilm detection. Sensors and Actuators, B: *Chemical* 2012; 163: 136-145.

[489] Wang Q, Zhang T. Review of mathematical models for biofilms. *Solid State Communications* 2010; 150: 1009-1022.

[490] Gjaltema A, Arts PAM, Van Loosdrecht MCM et al. Heterogeneity of biofilms in rotating annular reactors: Occurrence, structure, and consequences. *Biotechnology and Bioengineering* 1994; 44: 194-204.

[491] Zhang TC, Bishop PL. Density, porosity, and pore structure of biofilms. *Water Research* 1994; 28: 2267-2277.

[492] Picioreanu C, Van Loosdrecht MCM, Heijnen JJ. Discrete-differential modelling of biofilm structure. *Water Science and Technology* 1999; 39: 115-122.

[493] Jevšnik M, Ovca A, Bauer M et al. Food safety knowledge and practices among elderly in Slovenia. *Food Control* 2013; 31: 284-290.

[494] Jevšnik M, Hoyer S, Raspor P. Food safety knowledge and practices among pregnant and non-pregnant women in Slovenia. *Food Control* 2008; 19: 526-534.

[495] [Anonym]. CDC. *Vital signs: Incidence and trends of infection with pathogens transmitted commonly through food - foodborne diseases active surveillance network,* 10 US Sites, 1996-2010 2011;.

[496] Scallan E, Griffin PM, Angulo FJ et al. Foodborne illness acquired in the United states-Unspecified agents. *Emerging Infectious Diseases* 2011; 17: 16-22.

[497] Aarnisalo K, Lundén J, Korkeala H et al. Susceptibility of Listeria monocytogenes strains to disinfectants and chlorinated alkaline cleaners at cold temperatures. *LWT - Food Science and Technology* 2007; 40: 1041-1048.

[498] Dykes GA, Sampathkumar B, Korber DR. Planktonic or biofilm growth affects survival, hydrophobicity and protein expression patterns of a pathogenic Campylobacter jejuni strain. *International Journal of Food Microbiology* 2003; 89: 1-10.

[499] Lapidot A, Romling U, Yaron S. Biofilm formation and the survival of Salmonella Typhimurium on parsley. *International Journal of Food Microbiology* 2006; 109: 229-233.

[500] Bridier A, Sanchez-Vizuete P, Guilbaud M et al. Biofilm-associated persistence of food-borne pathogens. *Food Microbiology;*.

[501] Bridier A, Sanchez-Vizuete P, Guilbaud M et al. Biofilm-associated persistence of food-borne pathogens. *Food Microbiology* 2014;.

[502] Furukawa S, Akiyoshi Y, O'Toole GA et al. Sugar fatty acid esters inhibit biofilm formation by food-borne pathogenic bacteria. *International Journal of Food Microbiology* 2010; 138: 176-180.

[503] Janssens JCA, Steenackers H, Robijns S et al. Brominated furanones inhibit biofilm formation by Salmonella enterica serovar Typhimurium. *Applied and Environmental Microbiology* 2008; 74: 6639-6648.

[504] Al-Ahmad A, Maier J, Follo M et al. Food-borne Enterococci Integrate Into Oral Biofilm: An In Vivo Study. *Journal of Endodontics* 2010; 36: 1812-1819.

[505] Wang H-H, Ye K-P, Zhang Q-Q et al. Biofilm formation of meat-borne Salmonella enterica and inhibition by the cell-free supernatant from Pseudomonas aeruginosa. *Food Control* 2013; 32: 650-658.

[506] Bower CK, McGuire J, Daeschel MA. The adhesion and detachment of bacteria and spores on food-contact surfaces. *Trends in Food Science and Technology* 1996; 7: 152-157.

[507] Sidhu MS, Heir E, Sørum H et al. Genetic linkage between resistance to quaternary ammonium compounds and β-lactam antibiotics in food-

related Staphylococcus spp. *Microbial Drug Resistance* 2001; 7: 363-371.

[508] Sallam KI, Mohammed MA, Hassan MA et al. Prevalence, molecular identification and antimicrobial resistance profile of Salmonella serovars isolated from retail beef products in Mansoura, Egypt. *Food Control* 2014; 38: 209-214.

[509] Flemming HC, Ridgway H. Biofilm control: conventional and alternative approaches. *Marine and Industrial Biofouling* 2009; 4: 103-117.

[510] Storgårds E, Priha O. 16 - Biofilms and brewing. In: Fratamico PM, Annous BA, Gunther NW, Hrsg. *Biofilms in the Food and Beverage Industries: Woodhead Publishing;* 2009:432-454.

[511] Olajire AA. The brewing industry and environmental challenges. *Journal of Cleaner Production;.*

[512] Yamauchi Y, Okamoto T, Murayama H et al. Beer brewing using an immobilized yeast bioreactor design of an immobilized yeast bioreactor for rapid beer brewing system. *Journal of Fermentation and Bioengineering* 1994; 78: 443-449.

[513] Wirtanen G, Salo S. 3 - Biofilm risks. In: Lelieveld HLM, Mostert MA, Holah J, Hrsg. Handbook of Hygiene Control in the Food Industry: *Woodhead Publishing;* 2005:46-68.

[514] Shikongo-Nambabi M. Control of bacterial contamination during marine fish processing. *Journal of Biology* 2011; 3: 1-17.

[515] Elexson N, Afsah-Hejri L, Rukayadi Y et al. Effect of detergents as antibacterial agents on biofilm of antibiotics-resistant Vibrio parahaemolyticus isolates. *Food Control* 2014; 35: 378-385.

[516] Cruz CD, Fletcher GC. Prevalence and biofilm-forming ability of Listeria monocytogenes in New Zealand mussel (Perna canaliculus) processing plants. *Food Microbiology* 2011; 28: 1387-1393.

[517] Fallah AA, Saei-Dehkordi SS, Mahzounieh M. Occurrence and antibiotic resistance profiles of Listeria monocytogenes isolated from seafood products and market and processing environments in Iran. *Food Control* 2013; 34: 630-636.

[518] Skåra T, Rosnes JT, Leadley C. 4 - Microbial decontamination of seafood. In: Demirci A, Ngadi MO, Hrsg. *Microbial Decontamination in the Food Industry: Woodhead Publishing;* 2012:96-124.

[519] Valenzuela AS, Benomar N, Abriouel H et al. Isolation and identification of Enterococcus faecium from seafoods: Antimicrobial

resistance and production of bacteriocin-like substances. *Food Microbiology* 2010; 27: 955-961.

[520] Vázquez-Sánchez D, Cabo ML, Ibusquiza PS et al. Biofilm-forming ability and resistance to industrial disinfectants of Staphylococcus aureus isolated from fishery products. *Food Control* 2014; 39: 8-16.

[521] Hansson K, Andersson T, Skepö M. Adhesion of fermented diary products to packaging materials. Effect of material functionality, storage time, and fat content of the product. An empirical study. *Journal of Food Engineering* 2012; 111: 318-325.

[522] Uraz G, Gündüz ST. Investigation of the presence of biofilm in Bacillus subtilis, Bacillus licheniformis and Bacillus cereus which are isolated from raw milks. *Current Opinion in Biotechnology* 2013; 24, Supplement 1: S101-S102.

[523] Harvey J, Keenan KP, Gilmour A. Assessing biofilm formation by Listeria monocytogenes strains. *Food Microbiology* 2007; 24: 380-392.

[524] Giaouris E, Heir E, Hébraud M et al. Attachment and biofilm formation by foodborne bacteria in meat processing environments: Causes, implications, role of bacterial interactions and control by alternative novel methods. *Meat Science* 2014; 97: 298-309.

[525] Jahan M, Holley RA. Incidence of virulence factors in enterococci from raw and fermented meat and biofilm forming capacity at 25°C and 37°C. *International Journal of Food Microbiology* 2014; 170: 65-69.

[526] De Oliveira DCV, Fernandes Júnior A, Kaneno R et al. Ability of salmonella spp. to produce biofilm is dependent on temperature and surface material. *Foodborne Pathogens and Disease* 2014; 11: 478-483.

[527] Ban GH, Yoon H, Kang DH. A comparison of saturated steam and superheated steam for inactivation of Escherichia coli O157: H7, Salmonella Typhimurium, and Listeria monocytogenes biofilms on polyvinyl chloride and stainless steel. *Food Control* 2014; 40: 344-350.

[528] Gough NL, Dodd CER. The survival and disinfection of Salmonella typhimurium on chopping board surfaces of wood and plastic. *Food Control* 1998; 9: 363-368.

[529] Helke DM, Somers EB, Wong ACL. Attachment of Listeria monocytogenes and Salmonella typhimurium to stainless steel and Buna-N in the presence of milk and individual milk components. *J Food Prot* 1993; 56: 479-484.

[530] Prouty AM, Gunn JS. Comparative Analysis of Salmonella enterica Serovar Typhimurium Biofilm Formation on Gallstones and on Glass. *Infection and Immunity* 2003; 71: 7154-7158.

[531] Nguyen HDN, Yang YS, Yuk HG. Biofilm formation of Salmonella Typhimurium on stainless steel and acrylic surfaces as affected by temperature and pH level. *LWT - Food Science and Technology* 2014; 55: 383-388.

[532] Rivas L, Fegan N, Dykes GA. Attachment of Shiga toxigenic Escherichia coli to stainless steel. *International Journal of Food Microbiology* 2007; 115: 89-94.

[533] Posada-Izquierdo GD, Pérez-Rodríguez F, Zurera G. Mathematical quantification of microbial inactivation of Escherichia coli O157:H7 and Salmonella spp. on stainless steel surfaces soiled with different vegetable juice substrates. *Food Research International* 2013; 54: 1688-1698.

[534] Visvalingam J, Holley RA. Adherence of cold-adapted Escherichia coli O157:H7 to stainless steel and glass surfaces. *Food Control* 2013; 30: 575-579.

[535] Moritz MM, Flemming H-C, Wingender J. Integration of Pseudomonas aeruginosa and Legionella pneumophila in drinking water biofilms grown on domestic plumbing materials. *International Journal of Hygiene and Environmental Health* 2010; 213: 190-197.

[536] Cloete TE, Jacobs L. Surfactants and the attachment of Pseudomonas aeruginosa to 3CR12 stainless steel and glass. *Water SA* 2001; 27: 21-26.

[537] Bourion F, Cerf O. Disinfection efficacy against pure-culture and mixed-population biofilms of Listeria innocua and Pseudomonas aeruginosa on stainless steel, Teflon® and rubber. *Sciences des Aliments* 1996; 16: 151-166.

[538] Goh SG, Leili A-H, Kuan CH et al. Transmission of Listeria monocytogenes from raw chicken meat to cooked chicken meat through cutting boards. *Food Control* 2014; 37: 51-55.

[539] Adetunji VO, Isola TO. Crystal violet binding assay for assessment of biofilm formation by Listeria monocytogenes and Listeria spp on wood, steel and glass surfaces. *Global Veterinaria* 2011; 6: 6-10.

[540] Perni S, Aldsworth TG, Jordan SJ et al. The resistance to detachment of dairy strains of Listeria monocytogenes from stainless steel by shear stress is related to the fluid dynamic characteristics of the location of isolation. *International Journal of Food Microbiology* 2007; 116: 384-390.

[541] Li R, Kuda T, Yano T. Effect of food residues on efficiency of surfactant disinfectants against food related pathogens adhered on

polystyrene and ceramic surfaces. *LWT - Food Science and Technology* 2014; 57: 200-206.

[542] Hamadi F, Latrache H, Zekraoui M et al. Effect of pH on surface energy of glass and Teflon and theoretical prediction of Staphylococcus aureus adhesion. *Materials Science and Engineering: C* 2009; 29: 1302-1305.

[543] Klavenes A, Stalheim T, Sjøvold O et al. Attachment of Bacillus Cereus Spores With and Without Appendages to Stainless Steel Surfaces. *Food and Bioproducts Processing* 2002; 80: 312-318.

[544] Hornstra LM, de Leeuw PLA, Moezelaar R et al. Germination of Bacillus cereus spores adhered to stainless steel. *International Journal of Food Microbiology* 2007; 116: 367-371.

[545] Kumari S, Sarkar PK. In vitro model study for biofilm formation by Bacillus cereus in dairy chilling tanks and optimization of clean-in-place (CIP) regimes using response surface methodology. *Food Control* 2014; 36: 153-158.

[546] Nguyen VT, Turner MS, Dykes GA. Effect of temperature and contact time on campylobacter jejuni attachment to, and probability of detachment from, stainless steel. *Journal of Food Protection* 2010; 73: 832-838.

[547] Sanders SQ, Boothe DH, Frank JF et al. Culture and detection of Campylobacter jejuni within mixed microbial populations of biofilms on stainless steel. *Journal of Food Protection* 2007; 70: 1379-1385.

[548] Gunther Iv NW, Chen C-Y. The biofilm forming potential of bacterial species in the genus Campylobacter. *Food Microbiology* 2009; 26: 44-51.

[549] Nguyen VT, Turner MS, Dykes GA. Influence of cell surface hydrophobicity on attachment of Campylobacter to abiotic surfaces. *Food Microbiology* 2011; 28: 942-950.

[550] Whipps JM, Hand P, Pink DAC et al. Chapter 7 Human Pathogens and the Phyllosphere. *Advances in Applied Microbiology* 2008; 64: 183-221.

[551] Patel J, Singh M, Macarisin D et al. Differences in biofilm formation of produce and poultry Salmonella enterica isolates and their persistence on spinach plants. *Food Microbiology* 2013; 36: 388-394.

[552] Suslow TV. Water disinfection, a practical approach to calculating dose values for preharvest and postharvest applications. *Water disinfection: A practical approach to calculating dose values for preharvest and postharvest applications* 2001;.

[553] López-Gálvez F, Gil MI, Truchado P et al. Cross-contamination of fresh-cut lettuce after a short-term exposure during pre-washing cannot

be controlled after subsequent washing with chlorine dioxide or sodium hypochlorite. *Food Microbiology* 2010; 27: 199-204.

[554] Keskinen LA, Burke A, Annous BA. Efficacy of chlorine, acidic electrolyzed water and aqueous chlorine dioxide solutions to decontaminate Escherichia coli O157:H7 from lettuce leaves. *International Journal of Food Microbiology* 2009; 132: 134-140.

[555] Elhariry HM. Attachment strength and biofilm forming ability of Bacillus cereus on green-leafy vegetables: Cabbage and lettuce. *Food Microbiology* 2011; 28: 1266-1274.

[556] Bonsaglia ECR, Silva NCC, Fernades Júnior A et al. Production of biofilm by Listeria monocytogenes in different materials and temperatures. *Food Control* 2014; 35: 386-391.

[557] Caron C. CDC: *Cantaloupe Listeria outbreak deadliest in a decade* - ABC News 2011;.

[558] Neuman W. *Listeria outbreak traced to Colorado cantaloupe packing shed* - NYTimes 2011;.

[559] Sapers GM. Efficacy of Washing and Sanitizing Methods for Disinfection of Fresh Fruit and Vegetable Products. *Food Technology and Biotechnology* 2001; 39: 305-311.

[560] Liao CH, Cooke PH. Response to trisodium phosphate treatment of Salmonella Chester attached to fresh-cut green pepper slices. *Canadian Journal of Microbiology* 2001; 47: 25-32.

[561] Gandhi M, Golding S, Yaron S et al. Use of green fluorescent protein expressing Salmonella Stanley to investigate survival, spatial location, and control on alfalfa sprouts. *Journal of Food* Protection 2001; 64: 1891-1898.

[562] Han Y, Sherman DM, Linton RH et al. The effects of washing and chlorine dioxide gas on survival and attachment of Escherichia coli O157: H7 to green pepper surfaces. *Food Microbiology* 2000; 17: 521-533.

[563] Ölmez H, Temur SD. Effects of different sanitizing treatments on biofilms and attachment of Escherichia coli and Listeria monocytogenes on green leaf lettuce. *LWT - Food Science and Technology* 2010; 43: 964-970.

[564] Fransisca L, Zhou B, Park H et al. The Effect of Calcinated Calcium and Chlorine Treatments on Escherichia coli O157:H7 87-23 Population Reduction in Radish Sprouts. *Journal of Food Science* 2011; 76: M404-M412.

[565] Sapers GM, Sites JE. Efficacy of 1% hydrogen peroxide wash in decontaminating apples and cantaloupe melons. *Journal of Food Science* 2003; 68: 1793-1797.

[566] Jessen B, Lammert L. Biofilm and disinfection in meat processing plants. *International Biodeterioration and Biodegradation* 2003; 51: 265-269.

[567] Salo S, Ehavald H, Raaska L et al. Microbial surveys in Estonian dairies. LWT - *Food Science and Technology* 2006; 39: 460-471.

[568] Lehner A, Riedel K, Eberl L et al. Biofilm formation, extracellular polysaccharide production, and cell-to-cell signaling in various Enterobacter sakazakii strains: Aspects promoting environmental persistence. *Journal of Food Protection* 2005; 68: 2287-2294.

[569] Waak E, Tham W, Danielsson-Tham ML. Prevalence and fingerprinting of Listeria monocytogenes strains isolated from raw whole milk in farm bulk tanks and in dairy plant receiving tanks. *Applied and Environmental Microbiology* 2002; 68: 3366-3370.

[570] Latorre AA, Van Kessel JS, Karns JS et al. Biofilm in milking equipment on a dairy farm as a potential source of bulk tank milk contamination with Listeria monocytogenes. *Journal of Dairy Science* 2010; 93: 2792-2802.

[571] Tiwari U, Walsh D, Rivas L et al. Modelling the interaction of storage temperature, pH, and water activity on the growth behaviour of Listeria monocytogenes in raw and pasteurised semi-soft rind washed milk cheese during storage following ripening. *Food Control* 2014; 42: 248-256.

[572] Beale DJ, Morrison PD, Palombo EA. Detection of Listeria in milk using non-targeted metabolic profiling of Listeria monocytogenes: A proof-of-concept application. *Food Control* 2014; 42: 343-346.

[573] Quero GM, Santovito E, Visconti A et al. Quantitative detection of Listeria monocytogenes in raw milk and soft cheeses: Culture-independent versus liquid- and solid-based culture-dependent real time PCR approaches. *LWT - Food Science and Technology* 2014; 58: 11-20.

[574] Pearson LJ, Williamson JH, Turner SA et al. Peripartum infection with Streptococcus uberis but not coagulase-negative staphylococci reduced milk production in primiparous cows. *Journal of Dairy Science* 2013; 96: 158-164.

[575] Elias AO, Cortez A, Brandão PE et al. Molecular detection of Streptococcus agalactiae in bovine raw milk samples obtained directly from bulk tanks. *Research in Veterinary Science* 2012; 93: 34-38.

[576] Beecher C, Daly M, Ross RP et al. Characterization of the bovine innate immune response in milk somatic cells following intramammary infection with Streptococcus dysgalactiae subspecies dysgalactiae. *Journal of Dairy Science* 2012; 95: 5720-5729.

[577] Wiedmann M, Weilmeier D, Dineen SS et al. Molecular and phenotypic characterization of Pseudomonas spp. isolated from milk. *Applied and Environmental Microbiology* 2000; 66: 2085-2095.

[578] Cregenzán-Alberti O, Halpin RM, Whyte P et al. Suitability of ccRSM as a tool to predict inactivation and its kinetics for Escherichia coli, Staphylococcus aureus and Pseudomonas fluorescens in homogenized milk treated by manothermosonication (MTS). *Food Control* 2014; 39: 41-48.

[579] Marotta M, Ryan JT, Hickey RM. The predominant milk oligosaccharide 6′-sialyllactose reduces the internalisation of Pseudomonas aeruginosa in human pneumocytes. *Journal of Functional Foods* 2014; 6: 367-373.

[580] Van Tassell JA, Martin NH, Murphy SC et al. Evaluation of various selective media for the detection of Pseudomonas species in pasteurized milk. *Journal of Dairy Science* 2012; 95: 1568-1574.

[581] Holah JT, Lelieveld HLM, Napper D. 9 - Cleaning and disinfection practices in food processing. In Hygiene in Food Processing (Second edition): *Woodhead Publishing;* 2014:259-304.

[582] Shi X, Zhu X. Biofilm formation and food safety in food industries. *Trends in Food Science & Technology* 2009; 20: 407-413.

[583] Rajkowski KT. Biofilms in fish processing. *Biofilms in the food and beverage industries* 2009: 499-516.

[584] Faruque SM, Biswas K, Nashir Udden SM et al. Transmissibility of cholera: In vivo-formed biofilms and their relationship to infectivity and persistence in the environment. *Proceedings of the National Academy of Sciences of the United States of America* 2006; 103: 6350-6355.

[585] Enos-Berlage JL, Guvener ZT, Keenan CE et al. Genetic determinants of biofilm development of opaque and translucent Vibrio parahaemolyticus. *Molecular Microbiology* 2005; 55: 1160-1182.

[586] Joseph LA, Wright AC. Expression of Vibrio vulnificus Capsular Polysaccharide Inhibits Biofilm Formation. *Journal of Bacteriology* 2004; 186: 889-893.

[587] Momba MNB, Binda MA. Combining chlorination and chloramination processes for the inhibition of biofilm formation in drinking surface

water system models. *Journal of Applied Microbiology* 2002; 92: 641-648.

[588] Guobjörnsdóttir B, Einarsson H, Thorkelsson G. Microbial adhesion to processing lines for fish fillets and cooked shrimp: Influence of stainless steel surface finish and presence of gram-negative bacteria on the attachment of listeria monocytogenes. *Food Technology and Biotechnology* 2005; 43: 55-61.

[589] Gram L, Huss HH. Microbiological spoilage of fish and fish products. *International Journal of Food Microbiology* 1996; 33: 121-137.

[590] Vestby LK, Møretrø T, Langsrud S et al. Biofilm forming abilities of Salmonella are correlated with persistence in fish meal- and feed factories. *BMC Veterinary Research 2009; 5.*

[591] Arkoudelos JS, Samaras FJ, Tassou CC. Survival of Staphylococcus aureus and Salmonella Enteritidis on salted sardines (Sardina pilchardus) during ripening. *Journal of Food Protection* 2003; 66: 1479-1481.

[592] Atanassova V, Reich F, Klein G. Microbiological quality of Sushi from Sushi bars and retailers. *Journal of Food Protection* 2008; 71: 860-864.

[593] Guerin PJ, De Jong B, Heir E et al. Outbreak of Salmonella Livingstone infection in Norway and Sweden due to contaminated processed fish products. *Epidemiology and Infection* 2004; 132: 889-895.

[594] Brooks JD, Flint SH. Biofilms in the food industry: Problems and potential solutions. International *Journal of Food Science and Technology* 2008; 43: 2163-2176.

[595] Sala C, Morar A, Colibar O et al. Antibiotic resistance of gram negative bacteria isolated from meat surface biofilms. *Romanian Biotechnological* Letters 2012; 17: 7483-7492.

[596] Marouani-Gadri N, Augier G, Carpentier B. Characterization of bacterial strains isolated from a beef-processing plant following cleaning and disinfection - Influence of isolated strains on biofilm formation by Sakaï and EDL 933 E. coli O157:H7. *International Journal of Food Microbiology* 2009; 133: 62-67.

[597] Marin C, Hernandiz A, Lainez M. Biofilm development capacity of Salmonella strains isolated in poultry risk factors and their resistance against disinfectants. *Poultry Science* 2009; 88: 424-431.

[598] Park SH, Jarquin R, Hanning I et al. Detection of Salmonella spp. survival and virulence in poultry feed by targeting the hilA gene. *Journal of Applied Microbiology* 2011; 111: 426-432.

[599] Chia TWR, Goulter RM, McMeekin T et al. Attachment of different Salmonella serovars to materials commonly used in a poultry processing plant. *Food Microbiology* 2009; 26: 853-859.

[600] Abdallah FB, Chaieb K, Zmantar T et al. Adherence assays and slime production of Vibrio alginolyticus and Vibrio parahaemolyticus. *Ensaios de adesão e produção de muco por Vibrio alginolyticus e Vibrio parahaemolyticus* 2009; 40: 394-398.

[601] Er B, Demirhan B, Onurdağ FK et al. Antimicrobial and antibiofilm effects of selected food preservatives against salmonella spp. isolated from chicken samples. *Poultry Science* 2014; 93: 695-701.

[602] Chan KF, Tran HL, Kanenaka RY et al. Survival of Clinical and Poultry-Derived Isolates of Campylobacter jejuni at a Low Temperature (4°C). *Applied and Environmental Microbiology* 2001; 67: 4186-4191.

[603] Hanning I, Jarquin R, Slavik M. Campylobacter jejuni as a secondary colonizer of poultry biofilms. *Journal of Applied Microbiology* 2008; 105: 1199-1208.

[604] Van Houdt R, Michiels CW. Role of bacterial cell surface structures in Escherichia coli biofilm formation. *Research in Microbiology* 2005; 156: 626-633.

[605] Kim H, Ryu JH, Beuchat LR. Attachment of and biofilm formation by Enterobacter sakazakii on stainless steel and enteral feeding tubes. *Applied and Environmental Microbiology* 2006; 72: 5846-5856.

[606] Castonguay MH, van der Schaaf S, Koester W et al. Biofilm formation by Escherichia coli is stimulated by synergistic interactions and co-adhesion mechanisms with adherence-proficient bacteria. *Research in Microbiology 2006*; 157: 471-478.

[607] Klayman BJ, Volden PA, Stewart PS et al. Escherichia coli 0157:H7 requires colonizing partner to adhere and persist in a capillary flow cell. *Environmental Science and Technology* 2009; 43: 2105-2111.

[608] Simpson Beauchamp C, Dourou D, Geornaras I et al. Sanitizer efficacy against Escherichia coli O157:H7 biofilms on inadequately cleaned meat-contact surface materials. *Food Protection Trends* 2012; 32: 173-182.

[609] Habimana O, Heir E, Langsrud S et al. Enhanced surface colonization by escherichia coli 0157:H7 in biofilms formed by an acinetobacter calcoaceticus isolate from meat-processing environments. *Applied and Environmental Microbiology* 2010; 76: 4557-4559.

[610] Burmølle M, Webb JS, Rao D et al. Enhanced biofilm formation and increased resistance to antimicrobial agents and bacterial invasion are

caused by synergistic interactions in multispecies biofilms. *Applied and Environmental Microbiology* 2006; 72: 3916-3923.

[611] Nilsson LM, Yakovenko O, Tchesnokova V et al. The cysteine bond in the Escherichia coli FimH adhesin is critical for adhesion under flow conditions. *Molecular Microbiology* 2007; 65: 1158-1169.

[612] Sofos JN, Geornaras I. Overview of current meat hygiene and safety risks and summary of recent studies on biofilms, and control of Escherichia coli O157:H7 in nonintact, and Listeria monocytogenes in ready-to-eat, meat products. *Meat Science* 2010; 86: 2-14.

[613] Garrido V, Vitas AI, García-Jalón I. Survey of Listeria monocytogenes in ready-to-eat products: Prevalence by brands and retail establishments for exposure assessment of listeriosis in Northern Spain. *Food Control* 2009; 20: 986-991.

[614] Pinto AD, Novelleo L, Montemurro F et al. Occurrence of Listeria monocytogenes in ready-to-eat foods from supermarkets in Southern Italy. *New Microbiologica* 2010; 33: 249-252.

[615] Podkowik M, Bystroń J, Bania J. Genotypes, antibiotic resistance, and virulence factors of staphylococci from ready-to-eat food. *Foodborne Pathogens and Disease* 2012; 9: 91-93.

[616] Nyenje ME, Green E, Ndip RN. Biofilm formation and adherence characteristics of listeria ivanovii strains isolated from ready-to-eat foods in Alice, South Africa. *The Scientific World Journal* 2012; 2012.

[617] Keskinen LA, Todd ECD, Ryser ET. Transfer of surface-dried Listeria monocytogenes from stainless steel knife blades to roast Turkey breast. *Journal of Food Protection* 2008; 71: 176-181.

[618] Somers EB, Lee Wong AC. Efficacy of two cleaning and sanitizing combinations on Listeria monocytogenes biofilms formed at low temperature on a variety of materials in the presence of ready-to-eat meat residue. *Journal of Food Protection* 2004; 67: 2218-2229.

[619] Lindow SE, Brandl MT. Microbiology of the phyllosphere. *Applied and Environmental Microbiology* 2003; 69: 1875-1883.

[620] Morris CE, Monier JM, Jacques MA. Methods for observing microbial biofilms directly on leaf surfaces and recovering them for isolation of culturable microorganisms. *Applied and Environmental Microbiology* 1997; 63: 1570-1576.

[621] Fett WF, Cooke PH. A survey of native microbial aggregates on alfalfa, clover and mung bean sprout cotyledons for thickness as determined by confocal scanning laser microscopy. *Food Microbiology* 2005; 22: 253-259.

[622] Rudi K, Flateland SL, Hanssen JF et al. Development and evaluation of a 16S ribosomal DNA array-based approach for describing complex microbial communities in ready-to-eat vegetable salads packed in a modified atmosphere. *Applied and Environmental Microbiology* 2002; 68: 1146-1156.

[623] Burnett SL, Chen J, Beuchat LR. Attachment of Escherichia coli O157:H7 to the surfaces and internal structures of apples as detected by confocal scanning laser microscopy. *Applied and Environmental Microbiology* 2000; 66: 4679-4687.

[624] Niemira BA, Cooke PH. Escherichia coli O157:H7 Biofilm formation on romaine lettuce and spinach leaf surfaces reduces efficacy of irradiation and sodium hypochlorite washes. *Journal of Food Science* 2010; 75: M270-M277.

[625] Ells TC, Truelstrup Hansen L. Strain and growth temperature influence Listeria spp. attachment to intact and cut cabbage. *International Journal of Food Microbiology* 2006; 111: 34-42.

[626] Patel J, Sharma M. Differences in attachment of Salmonella enterica serovars to cabbage and lettuce leaves. *International Journal of Food Microbiology* 2010; 139: 41-47.

[627] Niemira BA, Solomon EB. Sensitivity of planktonic and biofilm-associated Salmonella spp. to ionizing radiation. *Applied and Environmental Microbiology* 2005; 71: 2732-2736.

[628] Warner JC, Rothwell SD, Keevil CW. Use of episcopic differential interference contrast microscopy to identify bacterial biofilms on salad leaves and track colonization by Salmonella Thompson. *Environmental Microbiology* 2008; 10: 918-925.

[629] Sant'Ana AS, Igarashi MC, Landgraf M et al. Prevalence, populations and pheno- and genotypic characteristics of Listeria monocytogenes isolated from ready-to-eat vegetables marketed in São Paulo, Brazil. *International Journal of Food Microbiology* 2012; 155: 1-9.

[630] Khalil RK, Frank JF. Behavior of escherichia coli 0157:H7 on damaged leaves of spinach, lettuce, cilantro, and parsley stored at abusive temperatures. Journal *of Food Protection* 2010; 73: 212-220.

[631] Faille C, Carpentier B. 11 - Food contact surfaces, surface soiling and biofilm formation. In: Fratamico PM, Annous BA, Gunther NW, Hrsg. *Biofilms in the Food and Beverage Industries:* Woodhead Publishing; 2009:303-330.

[632] Storgårds E, Simola H, Sjöberg AM et al. Hygiene of gasket materials used in food processing equipment part 1: New Materials. *Food and*

Bioproducts Processing: Transactions of the Institution of of Chemical Engineers, Part C 1999; 77: 137-145.

[633] Mariani C, Oulahal N, Chamba JF et al. Inhibition of Listeria monocytogenes by resident biofilms present on wooden shelves used for cheese ripening. *Food Control* 2011; 22: 1357-1362.

[634] Chaturongkasumrit Y, Takahashi H, Keeratipibul S et al. The effect of polyesterurethane belt surface roughness on Listeria monocytogenes biofilm formation and its cleaning efficiency. *Food Control* 2011; 22: 1893-1899.

[635] Howell D, Behrends B. A review of surface roughness in antifouling coatings illustrating the importance of cutoff length. *Biofouling* 2006; 22: 401-410.

[636] Scardino AJ, Harvey E, De Nys R. Testing attachment point theory: Diatom attachment on microtextured polyimide biomimics. *Biofouling* 2006; 22: 55-60.

[637] Bernardes PC, Andrade NJD, Penã WEL et al. Modeling of the adhesion of Bacillus cereus isolated from a dairy plant as a function of time and temperature. *Journal of Food Process Engineering* 2013; 36: 187-191.

[638] Arnold JW, Boothe DH, Suzuki O et al. Multiple imaging techniques demonstrate the manipulation of surfaces to reduce bacterial contamination and corrosion. *Journal of Microscopy* 2004; 216: 215-221.

[639] Boyd RD, Cole D, Rowe D et al. Cleanability of soiled stainless steel as studied by atomic force microscopy and time of flight secondary ion mass spectrometry. *Journal of Food Protection 2001*; 64: 87-93.

[640] Møretrø T, Hermansen L, Holck AL et al. Biofilm formation and the presence of the intercellular adhesion locus ica among staphylococi from food and food processing environments. *Applied and Environmental Microbiology* 2003; 69: 5648-5655.

[641] Borucki MK, Peppin JD, White D et al. Variation in Biofilm Formation among Strains of Listeria monocytogenes. *Applied and Environmental Microbiology* 2003; 69: 7336-7342.

[642] Midelet G, Carpentier B. Impact of cleaning and disinfection agents on biofilm structure and on microbial transfer to a solid model food. *Journal of Applied Microbiology* 2004; 97: 262-270.

[643] Olszewska MA. Microscopic findings for the study of biofilms in food environments. *Acta Biochimica Polonica* 2013; 60: 531-537.

[644] Ismaïl R, Aviat F, Michel V et al. Methods for recovering microorganisms from solid surfaces used in the food industry: A review

of the literature. *International Journal of Environmental Research and Public Health* 2013; 10: 6169-6183.

[645] Lindsay D, Brözel VS, Von Holy A. Spore formation in Bacillus subtilis biofilms. *Journal of Food Protection* 2005; 68: 860-865.

[646] Hall MR, McGillicuddy E, Kaplan LJ. Biofilm: Basic principles, pathophysiology, and implications for Clinicians. *Surgical Infections* 2014; 15: 1-7.

[647] Vickery K, Hu H, Jacombs AS et al. A review of bacterial biofilms and their role in device-associated infection. *Healthcare Infection* 2013; 18: 61-66.

[648] Harding JL, Reynolds MM. Combating medical device fouling. *Trends in Biotechnology* 2014; 32: 140-146.

[649] McCarty S, Woods E, Percival SL. Chapter Nine - Biofilms: From Concept to Reality. In: Percival SL, Williams DW, Randle J et al., Hrsg. *Biofilms in Infection Prevention and Control*. Boston: Academic Press; 2014:143-163.

[650] Henry GD, Wilson SK, Delk JR et al. Penile prosthesis cultures during revision surgery: A multicenter study. *Journal of Urology* 2004; 172: 153-156.

[651] Silverstein AD, Henry GD, Evans B et al. Biofilm Formation on Clinically Noninfected Penile Prostheses. *Journal of Urology* 2006; 176: 1008-1011.

[652] Arthur TD, Cavera VL, Chikindas ML. On bacteriocin delivery systems and potential applications. *Future Microbiology* 2014; 9: 235-248.

[653] Chaieb K, Mahdouani K, Bakhrouf A. Detection of icaA and icaD loci by polymerase chain reaction and biofilm formation by Staphylococcus epidermidis isolated from dialysate and needles in a dialysis unit. *Journal of Hospital Infection* 2005; 61: 225-230.

[654] Cerca N, Pier GB, Vilanova M et al. Quantitative analysis of adhesion and biofilm formation on hydrophilic and hydrophobic surfaces of clinical isolates of Staphylococcus epidermidis. *Research in Microbiology* 2005; 156: 506-514.

[655] Braceras I, Pacha-Olivenza MA, Calzado-Martín A et al. Decrease of Staphylococcal adhesion on surgical stainless steel after Si ion implantation. *Applied Surface Science* 2014; 310: 36-41.

[656] Braem A, Van Mellaert L, Mattheys T et al. Staphylococcal biofilm growth on smooth and porous titanium coatings for biomedical applications. *Journal of Biomedical Materials Research* - Part A 2014; 102: 215-224.

[657] Van Der Borden AJ, Van Der Mei HC, Busscher HJ. Electric block current induced detachment from surgical stainless steel and decreased viability of Staphylococcus epidermidis. *Biomaterials* 2005; 26: 6731-6735.

[658] Batalu D, Stanciuc AM, Moldovan L et al. Evaluation of pristine and Eu2O3-added MgB 2 ceramics for medical applications: Hardness, corrosion resistance, cytotoxicity and antibacterial activity. *Materials Science and Engineering* C 2014; 42: 350-361.

[659] Kaur S, Harjai K, Chhibber S. Bacteriophage mediated killing of Staphylococcus aureus in vitro on orthopaedic K wires in presence of linezolid prevents implant colonization. *PLoS ONE* 2014; 9.

[660] Kiedrowski MR, Kavanaugh JS, Malone CL et al. Nuclease modulates biofilm formation in community-associated methicillin-resistant staphylococcus aureus. *PLoS ONE* 2011; 6.

[661] Guembe M, Rodríguez-Créixems M, Martín-Rabadán P et al. The risk of catheter-related bloodstream infection after withdrawal of colonized catheters is low. *European Journal of Clinical Microbiology and Infectious Diseases* 2014; 33: 729-734.

[662] Tian Y, Cao H, Qiao Y et al. Antibacterial activity and cytocompatibility of titanium oxide coating modified by iron ion implantation. *Acta Biomaterialia* 2014;.

[663] Pradeep Kumar SS, Easwer HV, Maya Nandkumar A. Multiple Drug Resistant Bacterial Biofilms on Implanted Catheters - A Reservoir of Infection. *Journal of Association of Physicians of India 2013*; 61: 702-707.

[664] Shah JS, Venkatsurya PKC, Thein-Han WW et al. The role of nanocrystalline titania coating on nanostructured austenitic stainless steel in enhancing osteoblasts functions for regeneration of tissue. *Materials Science and Engineering* C 2011; 31: 458-471.

[665] Shi Z, Neoh KG, Zhong SP et al. In vitro antibacterial and cytotoxicity assay of multilayered polyelectrolyte-functionalized stainless steel. *Journal of Biomedical Materials Research* - Part A 2006; 76: 826-834.

[666] Hu SG, Jou CS, Yang MC. Effect of blending sodium polyethylene-5-sulfoisophthalate on adhesion of clinical bacteria on polyethylene terephthalate. *Journal of Applied Polymer Science* 2004; 91: 3587-3594.

[667] Shinde RB, Raut JS, Karuppayil MS. Biofilm formation by Candida albicans on various prosthetic materials and its fluconazole sensitivity: a kinetic study. *Mycoscience* 2012; 53: 220-226.

[668] Estivill D, Arias A, Torres-Lana A et al. Biofilm formation by five species of Candida on three clinical materials. *Journal of Microbiological Methods* 2011; 86: 238-242.

[669] Mishra M, Agrawal S, Raut S et al. Profile of yeasts isolated from urinary tracts of catheterized patients. *Journal of Clinical and Diagnostic Research* 2014; 8: 44-46.

[670] Raman N, Lee MR, Palecek SP et al. Polymer multilayers loaded with antifungal β-peptides kill planktonic Candida albicans and reduce formation of fungal biofilms on the surfaces of flexible catheter tubes. *Journal of Controlled Release* 2014;.

[671] Chassot F, Negri MFN, Svidzinski AE et al. Can intrauterine contraceptive devices be a Candida albicans reservoir? *Contraception* 2008; 77: 355-359.

[672] Tarquinio K, Confreda K, Shurko J et al. Activities of tobramycin and polymyxin E against Pseudomonas aeruginosa biofilm-coated medical grade endotracheal tubes. *Antimicrobial Agents and Chemotherapy* 2014; 58: 1723-1729.

[673] Loo CY, Young PM, Lee WH et al. Superhydrophobic, nanotextured polyvinyl chloride films for delaying Pseudomonas aeruginosa attachment to intubation tubes and medical plastics. *Acta Biomaterialia* 2012; 8: 1881-1890.

[674] Darouiche RO, Green G, Mansouri MD. Antimicrobial activity of antiseptic-coated orthopaedic devices. *International Journal of Antimicrobial Agents* 1998; 10: 83-86.

[675] Burnham GW, Cavanagh HD, Robertson DM. The impact of cellular debris on pseudomonas aeruginosa adherence to silicone hydrogel contact lenses and contact lens storage cases. *Eye and Contact Lens* 2012; 38: 7-15.

[676] Roosjen A, Busscher HJ, Norde W et al. Bacterial factors influencing adhesion of Pseudomonas aeruginosa strains to a poly(ethylene oxide) brush. *Microbiology* 2006; 152: 2673-2682.

[677] Chen-Charpentier BM, Stanescu D. Biofilm growth on medical implants with randomness. *Mathematical and Computer Modelling* 2011; 54: 1682-1686.

[678] Welliver Jr RC, Hanerhoff BL, Henry GD et al. Significance of Biofilm for the Prosthetic Surgeon. *Current Urology Reports* 2014; 15.

[679] McGee DC, Gould MK. Preventing complications of central venous catheterization. *New England Journal of Medicine* 2003; 348: 1123-1133.

[680] Tacconelli E, Tumbarello M, Pittiruti M et al. Central venous catheter-related sepsis in a cohort of 366 hospitalised patients. *European Journal of Clinical Microbiology and Infectious Diseases* 1997; 16: 203-209.
[681] Desrousseaux C, Sautou V, Descamps S et al. Modification of the surfaces of medical devices to prevent microbial adhesion and biofilm formation. *Journal of Hospital Infection* 2013; 85: 87-93.
[682] Rello J, Ollendorf DA, Oster G et al. Epidemiology and outcomes of ventilator-associated pneumonia in a large US database. *Chest* 2002; 122: 2115-2121.
[683] Raad II, Mohamed JA, Reitzel RA et al. The prevention of biofilm colonization by multidrug-resistant pathogens that cause ventilator-associated pneumonia with antimicrobial-coated endotracheal tubes. *Biomaterials* 2011; 32: 2689-2694.
[684] Thomsen TW, Setnik GS. Videos in clinical medicine. Male urethral catheterization. *The New England journal of medicine* 2006; 354.
[685] Hooton TM, Bradley SF, Cardenas DD et al. Diagnosis, prevention, and treatment of catheter-aassociated urinary tract infection in adults: 2009 international clinical practice guidelines from the infectious diseases society of America. *Clinical Infectious Diseases* 2010; 50: 625-663.
[686] Nicolle LE. Urinary Catheter-Associated Infections. *Infectious Disease Clinics of North America* 2012; 26: 13-27.
[687] Szczotka-Flynn LB, Pearlman E, Ghannoum M. Microbial contamination of contact lenses, lens care solutions, and their accessories: A literature review. *Eye and Contact Lens* 2010; 36: 116-129.
[688] Willcox MDP, Harmis NY, Holden BA. Bacterial populations on high-Dk silicone hydrogel contact lenses: Effect of length of wear in asymptomatic patients. *Clinical and Experimental Optometry* 2002; 85: 172-175.
[689] Cavanagh HD, Ladage PM, Li SL et al. Effects of daily and overnight wear of a novel hyper oxygen-transmissible soft contact lens on bacterial binding and corneal epithelium: A 13-month clinical trial. *Ophthalmolog*y 2002; 109: 1957-1969.
[690] Evans DJ, Fleiszig SMJ. Why does the healthy cornea resist pseudomonas aeruginosa infection? *American Journal of Ophthalmology* 2013; 155: 961-970.e962.
[691] Legeay G, Poncin-Epaillard F, Arciola CR. New surfaces with hydrophilic/hydrophobic characteristics in relations to (no)bioadhesion. *International Journal of Artificial Organs* 2006; 29: 453-461.

[692] Dutta D, Cole N, Willcox M. Factors influencing bacterial adhesion to contact lenses. *Molecular Vision* 2012; 18: 14-21.

[693] Wu P, Stapleton F, Willcox MD. The causes of and cures for contact lens-induced peripheral ulcer. *Eye Contact Lens* 2003; 29: S63-66; discussion S83-84, S192-194.

[694] Shousha MA, Santos ARC, Oechsler RA et al. A novel rat contact lens model for Fusarium keratitis. *Molecular Vision* 2013; 19: 2596-2605.

[695] Stapleton F, Carnt N. Contact lens-related microbial keratitis: How have epidemiology and genetics helped us with pathogenesis and prophylaxis. *Eye* 2012; 26: 185-193.

[696] Kaynak Onurdağ F, Özkan S, Özgen S et al. Candida albicans and Pseudomonas aeruginosa adhesion on soft contact lenses. *Graefe's Archive for Clinical and Experimental Ophthalmology* 2011; 249: 559-564.

[697] Lloyd AW, Faragher RGA, Denyer SP. Ocular biomaterials and implants. *Biomaterials* 2001; 22: 769-785.

[698] Kleinberg TT, Tzekov RT, Stein L et al. Vitreous Substitutes: A Comprehensive Review. *Survey of Ophthalmology* 2011; 56: 300-323.

[699] Kaliterna V, Kučišec-Tepeš N, Pejković L et al. An intrauterine device as a possible cause of change in the microbial flora of the female genital system. *Journal of Obstetrics and Gynaecology Research 2011; 37: 1035-1040.*

[700] Güdücü N, Gönenç G, Işçi H et al. Clinical importance of detection of bacterial vaginosis, trichomonas vaginalis, candida albicans and actinomyces in Papanicolaou smears. *Clinical and Experimental Obstetrics and Gynecology* 2012; 39: 333-336.

[701] Abdel-Hafeez M, El-Mehallaway N, Khalil I et al. Microbiological profile and biofilm formation on removed intrauterine contraceptive devices from a sample of Egyptian women. *Journal of Obstetrics and Gynaecology Research 2014*; 40: 1770-1776.

[702] Morrison CS, Turner AN, Jones LB. Highly effective contraception and acquisition of HIV and other sexually transmitted infections. *Best Practice and Research: Clinical Obstetrics and Gynaecology* 2009; 23: 263-284.

[703] Jonas L, Baguhl F, Wilken HP et al. Copper accumulation in actinomyces druses during endometritis after long-term use of an intrauterine contraceptive device. *Ultrastructural Pathology* 2002; 26: 323-329.

[704] Shanmughapriya S, Lency Francis A, Kavitha S et al. In vitro actinomycete biofilm development and inhibition by the polyene antibiotic, nystatin, on IUD copper surfaces. *Biofouling* 2012; 28: 929-935.

[705] Carrillo M, Valdez B, Vargas L et al. In vitro Actinomyces israelii biofilm development on IUD copper surfaces. *Contraception* 2010; 81: 261-264.

[706] Ducharne G, Girard J, Pasquier G et al. Hip prosthesis infection related to an unchecked intrauterine contraceptive device: A case report. *Orthopaedics and Traumatology: Surgery and Research* 2013; 99: 111-114.

[707] Farley TMM, Rosenberg MJ, Rowe PJ et al. Intrauterine devices and pelvic inflammatory disease: An international perspective. *Lancet* 1992; 339: 785-788.

[708] Matsuda K, Nakajima H, Khan KN et al. Preoperative diagnosis of pelvic actinomycosis by clinical cytology. *International Journal of Women's Health* 2012; 4: 527-533.

[709] Reichenbach J, Lopatin U, Mahlaoui N et al. Actinomyces in chronic granulomatous disease: An emerging and unanticipated pathogen. *Clinical Infectious Diseases* 2009; 49: 1703-1710.

[710] Lethaby A, Temple J, Santy J. *Pin site care for preventing infections associated with external bone fixators and pins.* Cochrane Database of Systematic Reviews 2008;.

[711] Hargreaves DG, Pajkos A, Deva AK et al. The role of biofilm formation in percutaneous kirschner-wire fixation of radial fractures. *The Journal of Hand Surgery: British & European Volume* 2002; 27: 365-368.

[712] Pacheco RJ, Saleh M. The role of external fixation in trauma. T*rauma* 2004; 6: 143-160.

[713] Lavini F, Donadelli A, Pizzoli A. Diaphyseal and metaphyseal fractures of the humerus. *Orthofix External Fixation in Trauma and Orthopaedics* 2000: 121-126.

[714] Cugola L, Atzei A. Diaphyseal fractures of the forearm. *Orthofix External Fixation in Trauma and Orthopaedics* 2000: 145-151.

[715] Timms A, Pugh H. Pin site care: guidance and key recommendations. *Nursing standard* (Royal College of Nursing (Great Britain): 1987) 2012; 27: 50-55; quiz 56.

[716] Bibbo C, Brueggeman J. Prevention and Management of Complications Arising from External Fixation Pin Sites. *Journal of Foot and Ankle Surgery* 2010; 49: 87-92.

[717] Kofman KE, Buckley T, McGrouther DA. Complications of transcutaneous metal devices. *European Journal of Plastic Surgery* 2012; 35: 673-682.

[718] Tillander J, Hagberg K, Hagberg L et al. Osseointegrated titanium implants for limb prostheses attachments: Infectious complications. *Clinical Orthopaedics and Related Research* 2010; 468: 2781-2788.

[719] Barraclough KA, Hawley CM, Playford EG et al. Prevention of access-related infection in dialysis. *Expert Review of Anti-Infective Therapy 2009*; 7: 1185-1200.

[720] Parker SL, Anderson WN, Lilienfeld S et al. Cerebrospinal shunt infection in patients receiving antibiotic-impregnated versus standard shunts: A review. *Journal of Neurosurgery: Pediatrics* 2011; 8: 259-265.

[721] Rose EA, Gelijns AC, Moskowitz AJ et al. Long-term use of a left ventricular assist device for end-stage heart failure. *New England Journal of Medicine* 2001; 345: 1435-1443.

[722] Pereda D, Conte JV. Left Ventricular Assist Device Driveline Infections. *Cardiology Clinics* 2011; 29: 515-527.

[723] Berman SJ, Johnson EW, Nakatsu C et al. Burden of infection in patients with end-stage renal disease requiring long-term dialysis. *Clinical Infectious Diseases* 2004; 39: 1747-1753.

[724] Lafrance JP, Rahme E, Lelorier J et al. Vascular Access-Related Infections: Definitions, Incidence Rates, and Risk Factors. *American Journal of Kidney Diseases* 2008; 52: 982-993.

[725] Jaber BL, Madias NE, Sarmento A et al. Bacterial infections in hemodialysis patients: Pathogenesis and prevention. *Kidney International* 2005; 67: 2508-2519.

[726] Attenello FJ, Garces-Ambrossi GL, Zaidi HA et al. Hospital costs associated with shunt infections in patients receiving antibiotic-impregnated shunt catheters versus standard shunt catheters. *Neurosurgery* 2010; 66: 284-289.

[727] Sbordone L, Bortolaia C. Oral microbial biofilms and plaque-related diseases: microbial communities and their role in the shift from oral health to disease. *Clinical oral investigations* 2003; 7: 181-188.

[728] Norowski Jr PA, Bumgardner JD. Biomaterial and antibiotic strategies for peri-implantitis. *Journal of Biomedical Materials Research - Part B Applied Biomaterials* 2009; 88: 530-543.

[729] Machtei EE, Frankenthal S, Levi G et al. Treatment of peri-implantitis using multiple applications of chlorhexidine chips: A double-blind,

randomized multi-centre clinical trial. *Journal of Clinical Periodontology* 2012; 39: 1198-1205.
[730] Zitzmann NU, Berglundh T. Definition and prevalence of peri-implant diseases. *Journal of Clinical Periodontology* 2008; 35: 286-291.
[731] Albrektsson T, Isidor F. Consensus report of session IV. *Proceedings of the 1st European Workshop on* Periodontology 1994: 365-369.
[732] Ten Cate JM. Biofilms, a new approach to the microbiology of dental plaque. *Odontology* 2006; 94: 1-9.
[733] Faust K, Sathirapongsasuti JF, Izard J et al. Microbial co-occurrence relationships in the Human Microbiome. *PLoS Computational Biology* 2012; 8.
[734] Romeo E, Ghisolfi M, Carmagnola D. Peri-implant diseases. A systematic review of the literature. *Minerva stomatologica* 2004; 53: 215-230.
[735] Lima EMCX, Koo H, Vacca Smith AM et al. Adsorption of salivary and serum proteins, and bacterial adherence on titanium and zirconia ceramic surfaces. *Clinical Oral Implants Research* 2008; 19: 780-785.
[736] Hahn R, Weiger R, Netuschil L et al. Microbial accumulation and vitality on different restorative materials. *Dental Materials* 1993; 9: 312-316.
[737] Auschill TM, Arweiler NB, Brecx M et al. The effect of dental restorative materials on dental biofilm. *European journal of oral sciences* 2002; 110: 48-53.
[738] Azevedo SMB, Kantorski KZ, Valandro LF et al. Effect of brushing with conventional versus whitening dentifrices on surface roughness and biofilm formation of dental ceramics. *General Dentistry* 2012; 60: e123-e130.
[739] Ono M, Nikaido T, Iked M et al. Surface properties of resin composite materials relative to biofilm formation. *Dental Materials Journal* 2007; 26: 613-622.
[740] Fúcio SBP, Carvalho FG, Sobrinho LC et al. The influence of 30-day-old Streptococcus mutans biofilm on the surface of esthetic restorative materials-An in vitro study. *Journal of Dentistry* 2008; 36: 833-839.
[741] Ready D, Qureshi F, Bedi R et al. Oral bacteria resistant to mercury and to antibiotics are present in children with no previous exposure to amalgam restorative materials. *FEMS Microbiology Letters* 2003; 223: 107-111.
[742] Pjetursson BE, Thoma D, Jung R et al. A systematic review of the survival and complication rates of implant-supported fixed dental

prostheses (FDPs) after a mean observation period of at least 5 years. *Clinical Oral Implants Research* 2012; 23: 22-38.

[743] Jung RE, Zembic A, Pjetursson BE et al. Systematic review of the survival rate and the incidence of biological, technical, and aesthetic complications of single crowns on implants reported in longitudinal studies with a mean follow-up of 5 years. *Clinical Oral Implants Research* 2012; 23: 2-21.

[744] Paquette DW, Ryan ME, Wilder RS. Locally delivered antimicrobials: clinical evidence and relevance. Journal of dental hygiene: JDH / *American Dental Hygienists' Association 2008*; 82 Suppl 3: 10-15.

[745] Trampuz A, Zimmerli W. Diagnosis and treatment of implant-associated septic arthritis and osteomyelitis. *Current Infectious Disease Reports* 2008; 10: 394-403.

[746] Trampuz A, Widmer AF. Infections associated with orthopedic implants. *Current Opinion in Infectious Diseases* 2006; 19: 349-356.

[747] Charlier C, Leclercq A, Cazenave B et al. Listeria monocytogenes-associated joint and bone infections: A study of 43 consecutive cases. *Clinical Infectious Diseases* 2012; 54: 240-248.

[748] Monksfield P, Chapple ILC, Matthews JB et al. Biofilm formation on bone-anchored hearing aids. *Journal of Laryngology and Otology* 2011; 125: 1125-1130.

[749] Lauderdale KJ, Malone CL, Boles BR et al. Biofilm dispersal of community-associated methicillin-resistant Staphylococcus aureus on orthopedic implant material. *Journal of Orthopaedic Research* 2010; 28: 55-61.

[750] Fang G, Keys TF, Gentry LO et al. Prosthetic valve endocarditis resulting from nosocomial bacteremia: A prospective, multicenter study. *Annals of Internal Medicine* 1993; 119: 560-567.

[751] Chamis AL, Peterson GE, Cabell CH et al. Staphylococcus aureus bacteremia in patients with permanent pacemakers or implantable cardioverter-defibrillators. *Circulation* 2001; 104: 1029-1033.

[752] Murdoch DR, Roberts SA, Fowler V.G, Jr. et al. Infection of orthopedic prostheses after Staphylococcus aureus bacteremia. *Clinical Infectious Diseases* 2001; 32: 647-649.

[753] Karchmer AW, Longworth DL. Infections of intracardiac devices. *Cardiology clinics* 2003; 21: 253-271, vii.

[754] Brabham WW, Wray DW, Gold MR. Microbiologic characteristics and antimicrobial susceptibility of pacemaker/ICD infections: A moving target! *Journal of Cardiovascular Electrophysiology* 2012; 23: 382-383.

[755] Jan E, Camou F, Texier-Maugein J et al. Microbiologic characteristics and in vitro susceptibility to antimicrobials in a large population of patients with cardiovascular implantable electronic device infection. *Journal of Cardiovascular Electrophysiology* 2012; 23: 375-381.

[756] Mond HG, Proclemer A. The 11th world survey of cardiac pacing and implantable cardioverter-defibrillators: Calendar year 2009 - A world society of Arrhythmia's project. *PACE - Pacing and Clinical Electrophysiology* 2011; 34: 1013-1027.

[757] Sohail MR, Uslan DZ, Khan AH et al. Management and Outcome of Permanent Pacemaker and Implantable Cardioverter-Defibrillator Infections. *Journal of the American College of Cardiology* 2007; 49: 1851-1859.

[758] Gandhi T, Crawford T, Riddell J. Cardiovascular Implantable Electronic Device Associated Infections. *Infectious Disease Clinics of North America* 2012; 26: 57-76.

[759] Arciola CR, Montanaro L, Costerton JW. New trends in diagnosis and control strategies for implant infections. *International Journal of Artificial Organs* 2011; 34: 727-736.

[760] Darouiche RO. Treatment of Infections Associated with Surgical Implants. *New England Journal of Medicine* 2004; 350: 1422-1429.

[761] Schmidt AH, Swiontkowski MF. Pathophysiology of infections after internal fixation of fractures. *The Journal of the American Academy of Orthopaedic Surgeons* 2000; 8: 285-291.

[762] Salwiczek M, Qu Y, Gardiner J et al. Emerging rules for effective antimicrobial coatings. *Trends in Biotechnology* 2014; 32: 82-90.

[763] Chen M, Yu Q, Sun H. Novel strategies for the prevention and treatment of biofilm related infections. *International Journal of Molecular Sciences* 2013; 14: 18488-18501.

[764] Basu B, Nath S. Fundamentals of biomaterials and biocompatibility. *Advanced Biomaterials* 2009: 1-18.

[765] Chen S, Li L, Zhao C et al. Surface hydration: Principles and applications toward low-fouling/nonfouling biomaterials. *Polymer* 2010; 51: 5283-5293.

[766] Li J, Kao WJ. Synthesis of polyethylene glycol (PEG) derivatives and PEGylated - Peptide biopolymer conjugates. *Biomacromolecules* 2003; 4: 1055-1067.

[767] Francolini I, Crisante F, Martinelli A et al. Synthesis of biomimetic segmented polyurethanes as antifouling biomaterials. *Acta Biomaterialia* 2012; 8: 549-558.

[768] Carlson GA, Dragoo JL, Samimi B et al. Bacteriostatic properties of biomatrices against common orthopaedic pathogens. *Biochemical and Biophysical Research Communications* 2004; 321: 472-478.

[769] Riedl CR, Witkowski M, Plas E et al. Heparin coating reduces encrustation of ureteral stents: A preliminary report. *International Journal of Antimicrobial Agents* 2002; 19: 507-510.

[770] Tenke P, Kovacs B, Jäckel M et al. The role of biofilm infection in urology. *World Journal of Urology* 2006; 24: 13-20.

[771] Shanks RMQ, Donegan NP, Graber ML et al. Heparin stimulates Staphylococcus aureus biofilm formation. *Infection and Immunity* 2005; 73: 4596-4606.

[772] Gao G, Lange D, Hilpert K et al. The biocompatibility and biofilm resistance of implant coatings based on hydrophilic polymer brushes conjugated with antimicrobial peptides. *Biomaterials* 2011; 32: 3899-3909.

[773] Andersen TE, Kingshott P, Palarasah Y et al. A flow chamber assay for quantitative evaluation of bacterial surface colonization used to investigate the influence of temperature and surface hydrophilicity on the biofilm forming capacity of uropathogenic Escherichia coli. *Journal of Microbiological Methods* 2010; 81: 135-140.

[774] Clston S, Lueptow RM, Gray KA. Controlling biofilm growth using reactive ceramic ultrafiltration membranes. *Journal of Membrane Science* 2009; 342: 263-268.

[775] Han JB, Wang X, Wang N et al. Effect of plasma treatment on hydrophilic properties of TiO2 thin films. *Surface and Coatings Technology* 2006; 200: 4876-4878.

[776] Triandafillu K, Balazs DJ, Aronsson BO et al. Adhesion of Pseudomonas aeruginosa strains to untreated and oxygen-plasma treated poly(vinyl chloride) (PVC) from endotracheal intubation devices. *Biomaterials* 2003; 24: 1507-1518.

[777] Gottenbos B, Grijpma DW, Van Der Mei HC et al. Antimicrobial effects of positively charged surfaces on adhering Gram-positive and Gram-negative bacteria. *Journal of Antimicrobial Chemotherapy* 2001; 48: 7-13.

[778] Terada A, Okuyama K, Nishikawa M et al. The effect of surface charge property on Escherichia coli initial adhesion and subsequent biofilm formation. *Biotechnology and Bioengineering* 2012; 109: 1745-1754.

[779] Epstein AK, Wong TS, Belisle RA et al. Liquid-infused structured surfaces with exceptional anti-biofouling performance. *Proceedings of*

the National Academy of Sciences of the United States of America 2012; 109: 13182-13187.

[780] Li J, Kleintschek T, Rieder A et al. Hydrophobic liquid-infused porous polymer surfaces for antibacterial applications. *ACS Applied Materials and Interfaces* 2013; 5: 6704-6711.

[781] Privett BJ, Youn J, Hong SA et al. Antibacterial fluorinated silica colloid superhydrophobic surfaces. *Langmuir* 2011; 27: 9597-9601.

[782] Lee ALZ, Ng VWL, Wang W et al. Block copolymer mixtures as antimicrobial hydrogels for biofilm eradication. *Biomaterials* 2013; 34: 10278-10286.

[783] Choi NC, Park SJ, Lee CG et al. Influence of surfactants on bacterial adhesion to metal oxide-coated surfaces. *Environmental Engineering Research* 2011; 16: 219-225.

[784] Zeraik AE, Nitschke M. Biosurfactants as agents to reduce adhesion of pathogenic bacteria to polystyrene surfaces: Effect of temperature and hydrophobicity. *Current Microbiology* 2010; 61: 554-559.

[785] Brambilla E, Ionescu A, Mazzoni A et al. Hydrophilicity of dentin bonding systems influences in vitro Streptococcus mutans biofilm formation. *Dental Materials* 2014;.

[786] Zhong LJ, Pang LQ, Che LM et al. Nafion coated stainless steel for anti-biofilm application. *Colloids and Surfaces B: Biointerfaces* 2013; 111: 252-256.

[787] Vani K, Thomas S, Prabhawathi V et al. In vitro biocompatiblity of modified polycarbonate as a biomaterial. *Colloids and Surfaces B: Biointerfaces* 2013; 108: 191-198.

[788] Ma C, Zhou H, Wu B et al. Preparation of polyurethane with zwitterionic side chains and their protein resistance. *ACS Applied Materials and Interfaces* 2011; 3: 455-461.

[789] Lowe AB, McCormick CL. Synthesis and solution properties of zwitterionic polymers. *Chemical Reviews* 2002; 102: 4177-4189.

[790] Siedenbiedel F, Tiller JC. Antimicrobial polymers in solution and on surfaces: Overview and functional principles. *Polymers* 2012; 4: 46-71.

[791] Timofeeva L, Kleshcheva N. Antimicrobial polymers: Mechanism of action, factors of activity, and applications. *Applied Microbiology and Biotechnology* 2011; 89: 475-492.

[792] Li G, Cheng G, Xue H et al. Ultra low fouling zwitterionic polymers with a biomimetic adhesive group. *Biomaterials* 2008; 29: 4592-4597.

[793] Chang Y, Chang WJ, Shih YJ et al. Zwitterionic sulfobetaine-grafted poly(VINYLIDENE FLUORIDE) membrane with highly effective

blood compatibility via atmospheric plasma-induced surface copolymerization. *ACS Applied Materials and Interfaces* 2011; 3: 1228-1237.

[794] Shen X, Zhao Y, Chen L. The construction of a zwitterionic PVDF membrane surface to improve biofouling resistance. *Biofouling* 2013; 29: 991-1003.

[795] Moore LE, Ledder RG, Gilbert P et al. In vitro study of the effect of cationic biocides on bacterial population dynamics and susceptibility. *Applied and Environmental Microbiology* 2008; 74: 4825-4834.

[796] Murata H, Koepsel RR, Matyjaszewski K et al. Permanent, non-leaching antibacterial surfaces-2: How high density cationic surfaces kill bacterial cells. *Biomaterials* 2007; 28: 4870-4879.

[797] Jayakumar R, Prabaharan M, Nair SV et al. Novel chitin and chitosan nanofibers in biomedical applications. *Biotechnology Advances* 2010; 28: 142-150.

[798] Rabea EI, Badawy MET, Stevens CV et al. Chitosan as antimicrobial agent: Applications and mode of action. *Biomacromolecules* 2003; 4: 1457-1465.

[799] Smith MJ, Adam G, Duncan HJ et al. The Effects of Cationic Surfactants on Marine Biofilm Growth on Hydrogels. *Estuarine, Coastal and Shelf Science* 2002; 55: 361-367.

[800] Williams DL, Haymond BS, Beck JP et al. In vivo efficacy of a silicone–cationic steroid antimicrobial coating to prevent implant-related infection. *Biomaterials* 2012; 33: 8641-8656.

[801] Ding X, Yang C, Lim TP et al. Antibacterial and antifouling catheter coatings using surface grafted PEG-b-cationic polycarbonate diblock copolymers. *Biomaterials* 2012; 33: 6593-6603.

[802] Zou P, Hartleb W, Lienkamp K. It takes walls and knights to defend a castle - Synthesis of surface coatings from antimicrobial and antibiofouling polymers. *Journal of Materials Chemistry* 2012; 22: 19579-19589.

[803] Cao B, Li L, Tang Q et al. The impact of structure on elasticity, switchability, stability and functionality of an all-in-one carboxybetaine elastomer. *Biomaterials* 2013; 34: 7592-7600.

[804] Waterhouse A, Wise SG, Yin Y et al. In vivo biocompatibility of a plasma-activated, coronary stent coating. *Biomaterials* 2012; 33: 7984-7992.

[805] Bilek MMM, Bax DV, Kondyurin A et al. Free radical functionalization of surfaces to prevent adverse responses to biomedical devices.

Proceedings of the National Academy of Sciences of the United States of America 2011; 108: 14405-14410.
[806] Wang BL, Ren KF, Chang H et al. Construction of degradable multilayer films for enhanced antibacterial properties. *ACS Applied Materials and Interfaces* 2013; 5: 4136-4143.
[807] Costa F, Carvalho IF, Montelaro RC et al. Covalent immobilization of antimicrobial peptides (AMPs) onto biomaterial surfaces. *Acta Biomaterialia* 2011; 7: 1431-1440.
[808] Forbes S, McBain AJ, Felton-Smith S et al. Comparative surface antimicrobial properties of synthetic biocides and novel human apolipoprotein E derived antimicrobial peptides. *Biomaterials* 2013; 34: 5453-5464.
[809] Héquet A, Humblot V, Berjeaud JM et al. Optimized grafting of antimicrobial peptides on stainless steel surface and biofilm resistance tests. *Colloids and Surfaces* B: Biointerfaces 2011; 84: 301-309.
[810] Anderson JM. Biological responses to materials. *Annual Review of Materials Science* 2001; 31: 81-110.
[811] Holmberg KV, Abdolhosseini M, Li Y et al. Bio-inspired stable antimicrobial peptide coatings for dental applications. *Acta Biomaterialia* 2013; 9: 8224-8231.
[812] Kazemzadeh-Narbat M, Lai BFL, Ding C et al. Multilayered coating on titanium for controlled release of antimicrobial peptides for the prevention of implant-associated infections. *Biomaterials* 2013; 34: 5969-5977.
[813] Shukla A, Fleming KE, Chuang HF et al. Controlling the release of peptide antimicrobial agents from surfaces. *Biomaterials* 2010; 31: 2348-2357.
[814] Quirynen M, Van der Mei HC, Bollen CM et al. The influence of surface-free energy on supra- and subgingival plaque microbiology. An in vivo study on implants. *Journal of Periodontology* 1994; 65: 162-167.
[815] Tang X, Flint SH, Bennett RJ et al. Biofilm growth of individual and dual strains of Klebsiella oxytoca from the dairy industry on ultrafiltration membranes. *Journal of Industrial Microbiology and Biotechnology* 2009; 36: 1491-1497.
[816] Litzler PY, Benard L, Barbier-Frebourg N et al. Biofilm formation on pyrolytic carbon heart valves: Influence of surface free energy, roughness, and bacterial species. *Journal of Thoracic and Cardiovascular Surgery* 2007; 134: 1025-1032.

[817] Kawai T, Saito K, Lee W. Protein binding to polymer brush, based on ion-exchange, hydrophobic, and affinity interactions. *Journal of Chromatography B: Analytical Technologies in the Biomedical and Life Sciences* 2003; 790: 131-142.

[818] Carlén A, Nikdel K, Wennerberg A et al. Surface characteristics and in vitro biofilm formation on glass ionomer and composite resin. *Biomaterials* 2001; 22: 481-487.

[819] Anselme K, Davidson P, Popa AM et al. The interaction of cells and bacteria with surfaces structured at the nanometre scale. *Acta Biomaterialia* 2010; 6: 3824-3846.

[820] Zhang F, Li H, Wang X et al. Hierarchically imprinted polymer substrates for enhanced attachment of Escherichia coli. *Journal of Colloid and Interface Science* 2010; 343: 109-114.

[821] Hochbaum AI, Aizenberg J. Bacteria pattern spontaneously on periodic nanostructure arrays. *Nano Letters* 2010; 10: 3717-3721.

[822] Carman ML, Estes TG, Feinberg AW et al. Engineered antifouling microtopographies - Correlating wettability with cell attachment. *Biofouling* 2006; 22: 11-21.

[823] Ercan B, Taylor E, Alpaslan E et al. Diameter of titanium nanotubes influences anti-bacterial efficacy. *Nanotechnology* 2011; 22.

[824] Puckett SD, Taylor E, Raimondo T et al. The relationship between the nanostructure of titanium surfaces and bacterial attachment. *Biomaterials* 2010; 31: 706-713.

[825] Yu WQ, Jiang XQ, Xu L et al. Osteogenic gene expression of canine bone marrow stromal cell and bacterial adhesion on titanium with different nanotubes. *Journal of Biomedical Materials Research - Part B Applied Biomaterials* 2011; 99 B: 207-216.

[826] Popat KC, Eltgroth M, LaTempa TJ et al. Decreased Staphylococcus epidermis adhesion and increased osteoblast functionality on antibiotic-loaded titania nanotubes. *Biomaterials* 2007; 28: 4880-4888.

[827] Zhao L, Chu PK, Zhang Y et al. Antibacterial coatings on titanium implants. *Journal of Biomedical Materials Research - Part B Applied Biomaterials* 2009; 91: 470-480.

[828] Fadeeva E, Truong VK, Stiesch M et al. Bacterial retention on superhydrophobic titanium surfaces fabricated by femtosecond laser ablation. *Langmuir* 2011; 27: 3012-3019.

[829] Loo C-Y, Young PM, Lee W-H et al. Superhydrophobic, nanotextured polyvinyl chloride films for delaying Pseudomonas aeruginosa

attachment to intubation tubes and medical plastics. *Acta Biomaterialia* 2012; 8: 1881-1890.

[830] Schumacher JF, Long CJ, Callow ME et al. Engineered nanoforce gradients for inhibition of settlement (attachment) of swimming algal spores. *Langmuir* 2008; 24: 4931-4937.

[831] Chung KK, Schumacher JF, Sampson EM et al. Impact of engineered surface microtopography on biofilm formation of Staphylococcus aureus. *Biointerphases* 2007; 2: 89-94.

[832] Epstein AK, Hochbaum AI, Kim P et al. Control of bacterial biofilm growth on surfaces by nanostructural mechanics and geometry. *Nanotechnology* 2011; 22.

[833] Neu HC. The crisis in antibiotic resistance. *Science* 1992; 257: 1064-1073.

[834] Stewart PS, Costerton JW. Antibiotic resistance of bacteria in biofilms. *Lancet* 2001; 358: 135-138.

[835] Hetrick EM, Schoenfisch MH. Reducing implant-related infections: Active release strategies. *Chemical Society Reviews* 2006; 35: 780-789.

[836] Charville GW, Hetrick EM, Geer CB et al. Reduced bacterial adhesion to fibrinogen-coated substrates via nitric oxide release. *Biomaterials* 2008; 29: 4039-4044.

[837] Lu Y, Slomberg DL, Sun B et al. Shape- and nitric oxide flux-dependent bactericidal activity of nitric oxide-releasing silica nanorods. *Small* 2013; 9: 2189-2198.

[838] Privett BJ, Broadnax AD, Bauman SJ et al. Examination of bacterial resistance to exogenous nitric oxide. *Nitric Oxide - Biology and Chemistry* 2012; 26: 169-173.

[839] Stefanini GG, Holmes Jr DR. Drug-eluting coronary-artery stents. *New England Journal of* Medicine 2013; 368: 254-265.

[840] Vasilev K, Poulter N, Martinek P et al. Controlled release of levofloxacin sandwiched between two plasma polymerized layers on a solid carrier. *ACS Applied Materials and Interfaces* 2011; 3: 4831-4836.

[841] Simovic S, Losic D, Vasilev K. Controlled drug release from porous materials by plasma polymer deposition. *Chemical Communications* 2010; 46: 1317-1319.

[842] Zhao B, Brittain WJ. Polymer brushes: Surface-immobilized macromolecules. *Progress in Polymer Science (Oxford)* 2000; 25: 677-710.

[843] Tiller JC, Liao CJ, Lewis K et al. Designing surfaces that kill bacteria on contact. *Proceedings of the National Academy of Sciences of the United States of America* 2001; 98: 5981-5985.

[844] Roosjen A, Van Der Mei HC, Busscher HJ et al. Microbial adhesion to poly(ethylene oxide) brushes: Influence of polymer chain length and temperature. *Langmuir* 2004; 20: 10949-10955.

[845] Roosjen A, Kaper HJ, van der Mei HC et al. Inhibition of adhesion of yeasts and bacteria by poly(ethylene oxide)-brushes on glass in a parallel plate flow chamber. *Microbiology* 2003; 149: 3239-3246.

[846] Kingshott P, Thissen H, Griesser HJ. Effects of cloud-point grafting, chain length, and density of PEG layers on competitive adsorption of ocular proteins. *Biomaterials* 2002; 23: 2043-2056.

[847] Kingshott P, Wei J, Bagge-Ravn D et al. Covalent attachment of poly(ethylene glycol) to surfaces, critical for reducing bacterial adhesion. *Langmuir* 2003; 19: 6912-6921.

[848] Park KD, Kim YS, Han DK et al. Bacterial adhesion on PEG modified polyurethane surfaces. *Biomaterials* 1998; 19: 851-859.

[849] Terada A, Yuasa A, Kushimoto T et al. Bacterial adhesion to and viability on positively charged polymer surfaces. *Microbiology* 2006; 152: 3575-3583.

[850] Kregiel D. Advances in biofilm control for food and beverage industry using organo-silane technology: A review. *Food Control* 2014; 40: 32-040.

[851] Arkles B. *Silane Coupling Agents: Connecting Across Boundaries* 2006;.

[852] Loontjens JA. Quaternary ammonium compounds. *Biomaterials associated infection* 2013: 379-404.

[853] Palermo EF, Lee DK, Ramamoorthy A et al. Role of cationic group structure in membrane binding and disruption by amphiphilic copolymers. *Journal of Physical Chemistry* B 2011; 115: 366-375.

[854] Gottenbos B, Van Der Mei HC, Klatter F et al. In vitro and in vivo antimicrobial activity of covalently coupled quaternary ammonium silane coatings on silicone rubber. *Biomaterials* 2002; 23: 1417-1423.

[855] Oosterhof JJH, Buijssen KJDA, Busscher HJ et al. Effects of quaternary ammonium silane coatings on mixed fungal and bacterial biofilms on tracheoesophageal shunt prostheses. *Applied and Environmental Microbiology* 2006; 72: 3673-3677.

[856] Gong SQ, Niu LN, Kemp LK et al. Quaternary ammonium silane-functionalized, methacrylate resin composition with antimicrobial

activities and self-repair potential. *Acta Biomaterialia* 2012; 8: 3270-3282.

[857] Kregiel D, Berlowska J, Mizerska U et al. Chemical modification of polyvinyl chloride and silicone elastomer in inhibiting adhesion of Aeromonas hydrophila. *World Journal of Microbiology and Biotechnology* 2013; 29: 1197-1206.

[858] Yoshino N, Sugaya S, Nakamura T et al. Synthesis and antimicrobial activity of squaternary ammonium silane coupling agents. *Journal of Oleo Science* 2011; 60: 429-438.

[859] Busalmen JP, De Sánchez SR. Electrochemical polarization-induced changes in the growth of individual cells and biofilms of Pseudomonas fluorescens (ATCC 17552). *Applied and Environmental Microbiology 2005*; 71: 6235-6240.

[860] Nakayama T, Wake H, Ozawa K et al. Use of a titanium nitride for electrochemical inactivation of marine bacteria. *Environmental Science and Technology* 1998; 32: 798-801.

[861] Van Der Borden AJ, Van Der Werf H, Van Der Mei HC et al. Electric current-induced detachment of Staphylococcus epidermidis biofilms from surgical stainless steel. *Applied and Environmental Microbiology* 2004; 70: 6871-6874.

[862] van der Borden AJ, Maathuis PGM, Engels E et al. Prevention of pin tract infection in external stainless steel fixator frames using electric current in a goat model. *Biomaterials* 2007; 28: 2122-2126.

[863] Nakasono S, Burgess JG, Takahashi K et al. Electrochemical prevention of marine biofouling with a carbon-chloroprene sheet. *Applied and Environmental Microbiology* 1993; 59: 3757-3762.

[864] Wake H, Takahashi H, Takimoto T et al. Development of an electrochemical antifouling system for seawater cooling pipelines of power plants using titanium. *Biotechnology and Bioengineering* 2006; 95: 468-473.

[865] Wagner VE, Koberstein JT, Bryers JD. Protein and bacterial fouling characteristics of peptide and antibody decorated surfaces of PEG-poly(acrylic acid) co-polymers. *Biomaterials* 2004; 25: 2247-2263.

[866] Chmielewski RAN, Frank JF. A predictive model for heat inactivation of Listeria monocytogenes biofilm on buna-N rubber. *LWT - Food Science and Technology* 2006; 39: 11-19.

[867] Gram L, Bagge-Ravn D, Ng YY et al. Influence of food soiling matrix on cleaning and disinfection efficiency on surface attached Listeria monocytogenes. *Food Control* 2007; 18: 1165-1171.

[868] Simões M, Simões LC, Machado I et al. Control of flow-generated biofilms with surfactants: Evidence of resistance and recovery. *Food and Bioproducts Processing* 2006; 84: 338-345.

[869] Pereira A, Mendes J, Melo LF. Using nanovibrations to monitor biofouling. *Biotechnology and Bioengineering* 2008; 99: 1407-1415.

[870] Philip-Chandy R, Scully PJ, Eldridge P et al. Optical fiber sensor for biofilm measurement using intensity modulation and image analysis. *IEEE Journal on Selected Topics in Quantum Electronics* 2000; 6: 764-772.

[871] Pereira A, Mendes J, Melo LF. Monitoring cleaning-in-place of shampoo films using nanovibration technology. *Sensors and Actuators, B: Chemical* 2009; 136: 376-382.

[872] Rogers J, Dowsett AB, Dennis PJ et al. Influence of plumbing materials on biofilm formation and growth of Legionella pneumophila in potable water systems. *Applied and Environmental Microbiology* 1994; 60: 1842-1851.

[873] Rogers J, Dowsett AB, Dennis PJ et al. Influence of temperature and plumbing material selection on biofilm formation and growth of Legionella pneumophila in a model potable water system containing complex microbial flora. *Applied and Environmental Microbiology* 1994; 60: 1585-1592.

[874] Griebe T, Flemming HC. Biocide-free antifouling strategy to protect RO membranes from biofouling. *Desalination* 1998; 118: 153-156.

[875] Frank JF, Chmielewski R. Influence of surface finish on the cleanability of stainless steel. *Journal of Food Protection* 2001; 64: 1178-1182.

[876] Meyer B. Approaches to prevention, removal and killing of biofilms. *International Biodeterioration and Biodegradation 2003*; 51: 249-253.

[877] Park SI, Daeschel MA, Zhao Y. Functional properties of antimicrobial lysozyme-chitosan composite films. *Journal of Food Science* 2004; 69: M215-M221.

[878] Weng YM, Chen MJ, Wenlung C. Antimicrobial food packaging materials from poly(ethylene-co-methacrylic acid). *LWT - Food Science and Technology* 1999; 32: 191-195.

[879] Thouvenin M, Langlois V, Briandet R et al. Study of erodable paint properties involved in antifouling activity. *Biofouling* 2003; 19: 177-186.

[880] Tsibouklis J, Stone M, A. Thorpe A et al. Inhibiting bacterial adhesion onto surfaces: The non-stick coating approach. *International Journal of Adhesion and Adhesives* 2000; 20: 91-96.

[881] Rosmaninho R, Santos O, Nylander T et al. Modified stainless steel surfaces targeted to reduce fouling - Evaluation of fouling by milk components. *Journal of Food Engineering* 2007; 80: 1176-1187.

[882] Hashimoto H. Evaluation of the anti-biofilm effect of a new anti-bacterial silver citrate/lecithin coating in an in-vitro experimental system using a modified Robbins device. Kansenshogaku zasshi *The Journal of the Japanese Association for Infectious Diseases* 2001; 75: 678-685.

[883] Klueh U, Wagner V, Kelly S et al. Efficacy of silver-coated fabric to prevent bacterial colonization and subsequent device-based biofilm formation. *Journal of Biomedical Materials Research* 2000; 53: 621-631.

[884] Uno T, Ohashi Y, Imayasu M. Antimicrobial efficacy tests of multipurpose contact lens care solutions simulating poor contact lens hygiene behaviors. *Eye and Contact Lens* 2012; 38: 388-393.

[885] Wu YT, Zhu H, Willcox M et al. Removal of biofilm from contact lens storage cases. *Investigative Ophthalmology and Visual Science* 2010; 51: 6329-6333.

[886] Willcox MDP, Zhu H, Vijay AK. Effect of a warming device on contact lens case contamination. *Eye and Contact Lens* 2012; 38: 394-399.

[887] Retuerto MA, Szczotka-Flynn L, Ho D et al. Efficacy of care solutions against contact lens-associated fusarium biofilms. *Optometry and Vision Science* 2012; 89: 382-391.

[888] Hall BJ, Jones L. Contact lens cases: The missing link in contact lens safety? *Eye and Contact Lens* 2010; 36: 101-105.

[889] Meylheuc T, Renault M, Bellon-Fontaine MN. Adsorption of a biosurfactant on surfaces to enhance the disinfection of surfaces contaminated with Listeria monocytogenes. *International Journal of Food Microbiology* 2006; 109: 71-78.

[890] Pereira MO, Machado S, Simões LC et al. Influence of material type and surface benzalkonium chloride preconditioning on biofilm formation and activity. *Proceedings of the international conference fouling, cleaning and disinfection in food processing* 2006: 347-353.

[891] Splendiani A, Livingston AG, Nicolella C. Control of membrane-attached biofilms using surfactants. *Biotechnology and Bioengineering* 2006; 94: 15-23.

[892] Denyer SP, Gorman SP, Sussman M. *Microbial Biofilms: Formation and Control* 1993;.

[893] Holah JT. Industrial monitoring: Hygiene in food processing. *Biofilms - Science and Technology* 1992: 645-659.

[894] Jacquelin LF, Le Magrex E, Brisset L et al. Synergic effect of enzymes or surfactants in association with a phenolic disinfectant on a bacterial biofilm. Synergie de l'association d'enzymes ou de surfactants et d'un desinfectant phenolique sur un biofilm *bacterien* 1994; 42: 425-431.
[895] Frank JF, Ehlers J, Wicker L. Removal of Listeria monocytogenes and poultry soil-containing biofilms using chemical cleaning and sanitizing agents under static conditions. *Food Prot Trends* 2003; 23: 654-663.
[896] Marques SC, Rezende JDGOS, Alves LADF et al. Formation of biofilms by Staphylococcus aureus on stainless steel and glass surfaces and its resistance to some selected chemical sanitizers. *Brazilian Journal of Microbiology* 2007; 38: 538-543.
[897] Salvia ACRD, Teodoro GR, Balducci I et al. Effectiveness of 2% peracetic acid for the disinfection of gutta-percha cones. *Brazilian Oral Research* 2011; 25: 23-27.
[898] Toté K, Horemans T, Vanden Berghe D et al. Inhibitory effect of biocides on the viable masses and matrices of Staphylococcus aureus and Pseudomonas aeruginosa biofilms. *Applied and Environmental Microbiology* 2010; 76: 3135-3142.
[899] Carmona D, Lalueza P, Balas F et al. Mesoporous silica loaded with peracetic acid and silver nanoparticles as a dual effect, highly efficient bactericidal agent. *Microporous and Mesoporous Materials* 2012; 161: 84-90.
[900] Lalueza P, Carmona D, Monzón M et al. Strong bactericidal synergy between peracetic acid and silver-exchanged zeolites. *Microporous and Mesoporous Materials* 2012; 156: 171-175.
[901] Martín-Espada MC, D'Ors A, Bartolomé MC et al. Peracetic acid disinfectant efficacy against Pseudomonas aeruginosa biofilms on polystyrene surfaces and comparison between methods to measure it. *LWT - Food Science and Technology* 2014; 56: 58-61.
[902] Park S-H, Cheon H-L, Park K-H et al. Inactivation of biofilm cells of foodborne pathogen by aerosolized sanitizers. *International Journal of Food Microbiology* 2012; 154: 130-134.
[903] Królasik J, Zakowska Z, Krępska M et al. Resistance of bacterial biofilms formed on stainless steel surface to disinfecting agent. *Polish Journal of Microbiology* 2010; 59: 281-287.
[904] Rossoni EMM, Gaylarde CC. Comparison of sodium hypochlorite and peracetic acid as sanitising agents for stainless steel food processing surfaces using epifluorescence microscopy. *International Journal of Food* Microbiology 2000; 61: 81-85.

[905] Exner M, Tuschewitzki GJ, Scharnagel J. Influence of biofilms by chemical disinfectants and mechanical cleaning. Zentralblatt fur Bakteriologie Mikrobiologie und Hygiene - *Abt 1 Orig B Hygiene* 1987; 183: 549-563.

[906] Mosteller TM, Bishop JR. Sanitizer efficacy against attached bacteria in a milk biofilm. *J Food Prot* 1993; 56: 34-41.

[907] Horvath M, Bilitzky L, Huttner J. Ozone 1985..

[908] Khadre MA, Yousef AE, Kim JG. Microbiological aspects of ozone applications in food: A review. *Journal of Food Science* 2001; 66: 1242-1252.

[909] Pascual A, Llorca I, Canut A. Use of ozone in food industries for reducing the environmental impact of cleaning and disinfection activities. *Trends in Food Science and Technology* 2007; 18: S29-S35.

[910] Dosti B, Guzel-Seydim Z, Greene AK. Effectiveness of ozone, heat and chlorine for destroying common food spoilage bacteria in synthetic media and biofilms. *International Journal of Dairy Technology* 2005; 58: 19-24.

[911] Lagrange F, Reiprich W, Hoffmann M. CIP-cleaning and desinfection with ozoned water. *CIP-reinigung und desinfektion mit ozonwasser* 2004; 84: 112-114.

[912] Tachikawa M, Yamanaka K, Nakamuro K. Studies on the disinfection and removal of biofilms by ozone water using an artificial microbial biofilm system. *Ozone: Science and Engineering 2009*; 31: 3-9.

[913] Di Ciccio P, Ghidini S, Zanardi E et al. Effects of gaseous ozone on food-borne pathogens. *Italian Journal of Food Science* 2014; 26: 116-118.

[914] Guzel-Seydim ZB, Greene AK, Seydim AC. Use of ozone in the food industry. *LWT - Food Science and Technology* 2004; 37: 453-460.

[915] Hanley-Onken E, Cohen N. The efficacy of ozonated water in biofilm control in USP purified water circulation and storage. *Pharmaceutical Engineering* 2013; 33: 40-50.

[916] Guzzon R, Nardin T, Micheletti O et al. Antimicrobial activity of ozone. Effectiveness against the main wine spoilage microorganisms and evaluation of impact on simple phenols in wine. *Australian Journal of Grape and Wine Research* 2013; 19: 180-188.

[917] Ozdemir HO, Buzoglu HD, Calt S et al. Effect of Ethylenediaminetetraacetic Acid and Sodium Hypochlorite Irrigation on Enterococcus faecalis Biofilm Colonization in Young and Old Human

Root Canal Dentin: In Vitro Study. *Journal of Endodontics* 2010; 36: 842-846.
[918] Da Silva PMB, Acosta EJTR, de Rezende Pinto L et al. Microscopical analysis of Candida albicans biofilms on heat-polymerised acrylic resin after chlorhexidine gluconate and sodium hypochlorite treatments. *Mycoses* 2011; 54: e712-e717.
[919] Spratt DA, Pratten J, Wilson M et al. An in vitro evaluation of the antimicrobial efficacy of irrigants on biofilms of root canal isolates. *International Endodontic Journal* 2001; 34: 300-307.
[920] Araújo P, Lemos M, Mergulhão F et al. Antimicrobial resistance to disinfectants in biofilms. *Science Against Microbial Pathogens: Communicating Current Research and Technological Advances* 2011: 826-834.
[921] Lomander A, Schreuders P, Russek-Cohen E et al. Evaluation of chlorines' impact on biofilms on scratched stainless steel surfaces. *Bioresource Technology* 2004; 94: 275-283.
[922] Simon FX, Berdalet E, Gracia FA et al. Seawater disinfection by chlorine dioxide and sodium hypochlorite. A comparison of biofilm formation. *Water, Air, and Soil Pollution* 2014; 225.
[923] Ordinola-Zapata R, Bramante CM, Aprecio RM et al. Biofilm removal by 6% sodium hypochlorite activated by different irrigation techniques. *International Endodontic Journal* 2014; 47: 659-666.
[924] Zand V, Milani AS, Amini M et al. Antimicrobial efficacy of photodynamic therapy and sodium hypochlorite on monoculture biofilms of enterococcus faecalis at different stages of development. *Photomedicine and Laser Surgery* 2014; 32: 245-251.
[925] Norwood DE, Gilmour A. The growth and resistance to sodium hypochlorite of Listeria monocytogenes in a steady-state multispecies biofilm. *Journal of Applied Microbiology* 2000; 88: 512-520.
[926] Liu H, Wei X, Ling J et al. Biofilm Formation Capability of Enterococcus faecalis Cells in Starvation Phase and Its Susceptibility to Sodium Hypochlorite. *Journal of Endodontics* 2010; 36: 630-635.
[927] Capita R, Riesco-Peláez F, Alonso-Hernando A et al. Exposure of Escherichia coli ATCC 12806 to sublethal concentrations of food-grade biocides influences its ability to form biofilm, resistance to antimicrobials, and ultrastructure. *Applied and Environmental Microbiology* 2014; 80: 1268-1280.

[928] Christina T, Penna V, Mazzola PG et al. The efficacy of chemical agents in cleaning and disinfection programs. *BMC Infectious Deseases* 2001; 1.

[929] Rideout K, Teschke K, Dimich-Ward H et al. Considering risks to healthcare workers from glutaraldehyde alternatives in high-level disinfection. *Journal of Hospital Infection* 2005; 59: 4-11.

[930] Kim J, Pitts B, Stewart PS et al. Comparison of the antimicrobial effects of chlorine, silver ion, and tobramycin on biofilm. *Antimicrobial Agents and Chemotherapy* 2008; 52: 1446-1453.

[931] Shikongo-Nambabi MNNN, Kachigunda B, Venter SN. Evaluation of oxidising disinfectants to control vibrio biofilms in treated seawater used for fish processing. *Water SA* 2010; 36: 215-220.

[932] Robbins JB, Fisher CW, Moltz AG et al. Elimination of Listeria monocytogenes biofilms by ozone, chlorine, and hydrogen peroxide. *Journal of Food Protection* 2005; 68: 494-498.

[933] Stojicic S, Shen Y, Haapasalo M. Effect of the source of biofilm bacteria, level of biofilm maturation, and type of disinfecting agent on the susceptibility of biofilm bacteria to antibacterial agents. *Journal of Endodontics* 2013; 39: 473-477.

[934] Perumal PK, Wand ME, Sutton JM et al. Evaluation of the effectiveness of hydrogen-peroxide-based disinfectants on biofilms formed by Gram-negative pathogens. *Journal of Hospital Infection* 2014;.

[935] Chaieb K, Zmantar T, Souiden Y et al. XTT assay for evaluating the effect of alcohols, hydrogen peroxide and benzalkonium chloride on biofilm formation of Staphylococcus epidermidis. *Microbial Pathogenesis* 2011; 50: 1-5.

[936] Gray NF. Chapter Thirty-Two - Chlorine Dioxide. In: Percival SL, Yates MV, Williams DW et al., Hrsg. *Microbiology of Waterborne Diseases* (Second Edition). London: Academic Press; 2014:591-598.

[937] Zhou SW, Xu FD, Li SM et al. Major origin of mutagenicity of chlorinated drinking water in China: Humic acid or pollutants. *Science of the Total Environment* 1997; 196: 191-196.

[938] Chen X, Stewart PS. Chlorine penetration into artificial biofilm is limited by a reaction-diffusion interaction. *Environmental Science and Technology* 1996; 30: 2078-2083.

[939] Xu X, Stewart PS, Chen X. Transport limitation of chlorine disinfection of Pseudomonas aeruginosa entrapped in alginate beads. *Biotechnology and Bioengineering* 1996; 49: 93-100.

[940] Scher K, Romling U, Yaron S. Effect of heat, acidification, and chlorination on Salmonella enterica serovar typhimurium cells in a biofilm formed at the air-liquid interface. *Applied and Environmental Microbiology* 2005; 71: 1163-1168.

[941] Joseph B, Otta SK, Karunasagar I. Biofilm formation by Salmonella spp. on food contact surfaces and their sensitivity to sanitizers. *International Journal of Food Microbiology* 2001; 64: 367-372.

[942] Nam H, Seo H-S, Bang J et al. Efficacy of gaseous chlorine dioxide in inactivating Bacillus cereus spores attached to and in a biofilm on stainless steel. *International Journal of Food Microbiology* 2014;.

[943] Schlisselberg DB, Yaron S. The effects of stainless steel finish on Salmonella Typhimurium attachment, biofilm formation and sensitivity to chlorine. *Food Microbiology* 2013; 35: 65-72.

[944] Trinetta V, Vaid R, Xu Q et al. Inactivation of Listeria monocytogenes on ready-to-eat food processing equipment by chlorine dioxide gas. *Food Control* 2012; 26: 357-362.

[945] Gao H, Liu C. Biochemical and morphological alteration of Listeria monocytogenes under environmental stress caused by chloramine-T and sodium hypochlorite. *Food Control* 2014; 46: 455-461.

[946] O'Leary D, Cabe EMM, McCusker MP et al. Microbiological study of biofilm formation in isolates of Salmonella enterica Typhimurium DT104 and DT104b cultured from the modern pork chain. *International Journal of Food Microbiology* 2013; 161: 36-43.

[947] Todd ECD, Michaels BS, Holah J et al. Outbreaks where food workers have been implicated in the spread of foodborne disease. Part 10. Alcohol-based antiseptics for hand disinfection and a comparison of their effectiveness with soaps. *Journal of Food* Protection 2010; 73: 2128-2140.

[948] Kampf G, Kramer A. Epidemiologic background of hand hygiene and evaluation of the most important agents for scrubs and rubs. *Clinical Microbiology Reviews* 2004; 17: 863-893.

[949] Dixit A, Pandey P, Mahajan R et al. Alcohol based hand sanitizers: Assurance and apprehensions revisited. Research Journal of Pharmaceutical, *Biological and Chemical Sciences* 2014; 5: 558-563.

[950] Møretrø T, Vestby LK, Nesse LL et al. Evaluation of efficacy of disinfectants against Salmonella from the feed industry. *Journal of Applied Microbiology* 2009; 106: 1005-1012.

[951] Milisavljevic V, Tran LP, Batmalle C et al. Benzyl alcohol and ethanol can enhance the pathogenic potential of clinical Staphylococcus

epidermidis strains. *American Journal of Infection Contro*l 2008; 36: 552-558.

[952] Chauhan NM, Raut JS, Mohan Karuppayil S. Acetaldehyde inhibits the yeast-to-hypha conversion and biofilm formation in Candida albicans. *Mycoscience* 2011; 52: 356-360.

[953] Walker JT, Bradshaw DJ, Fulford MR et al. Microbiological evaluation of a range of disinfectant products to control mixed-species biofilm contamination in a laboratory model of a dental unit water system. *Applied and Environmental Microbiology* 2003; 69: 3327-3332.

[954] Korem M, Gov Y, Rosenberg M. Global gene expression in Staphylococcus aureus following exposure to alcohol. *Microbial Pathogenesis* 2010; 48: 74-84.

[955] Stobie N, Duffy B, McCormack DE et al. Prevention of Staphylococcus epidermidis biofilm formation using a low-temperature processed silver-doped phenyltriethoxysilane sol-gel coating. *Biomaterials* 2008; 29: 963-969.

[956] Melaiye A, Youngs WJ. Silver and its application as an antimicrobial agent. *Expert Opinion on Therapeutic Patents* 2005; 15: 125-130.

[957] Panáček A, Kvítek L, Prucek R et al. Silver colloid nanoparticles: Synthesis, characterization, and their antibacterial activity. *Journal of Physical Chemistry* B 2006; 110: 16248-16253.

[958] Rupp ME, Fitzgerald T, Marion N et al. Effect of silver-coated urinary catheters: Efficacy, cost-effectiveness, and antimicrobial resistance. *American Journal of Infection Control* 2004; 32: 445-450.

[959] Taylor PL, Ussher AL, Burrell RE. Impact of heat on nanocrystalline silver dressings: Part I: Chemical and biological properties. *Biomaterials* 2005; 26: 7221-7229.

[960] Furno F, Morley KS, Wong B et al. Silver nanoparticles and polymeric medical devices: A new approach to prevention of infection? *Journal of Antimicrobial Chemotherapy* 2004; 54: 1019-1024.

[961] Knetsch MLW, Koole LH. New strategies in the development of antimicrobial coatings: The example of increasing usage of silver and silver nanoparticles. *Polymers* 2011; 3: 340-366.

[962] Kawahara K, Tsuruda K, Morishita M et al. Antibacterial effect of silver-zeolite on oral bacteria under anaerobic conditions. *Dental Materials* 2000; 16: 452-455.

[963] Baker C, Pradhan A, Pakstis L et al. Synthesis and antibacterial properties of silver nanoparticles. *Journal of Nanoscience and Nanotechnology* 2005; 5: 244-249.

[964] Kong H, Jang J. Antibacterial properties of novel poly(methyl methacrylate) nanofiber containing silver nanoparticles. *Langmuir* 2008; 24: 2051-2056.

[965] Harrison JJ, Ceri H, Yerly J et al. Metal ions may suppress or enhance cellular differentiation in Candida albicans and Candida tropicalis biofilms. *Applied and Environmental Microbiology* 2007; 73: 4940-4949.

[966] Monteiro DR, Gorup LF, Takamiya AS et al. The growing importance of materials that prevent microbial adhesion: antimicrobial effect of medical devices containing silver. *International Journal of Antimicrobial Agents* 2009; 34: 103-110.

[967] Tabrez Khan S, Ahamed M, Al-Khedhairy A et al. Biocidal effect of copper and zinc oxide nanoparticles on human oral microbiome and biofilm formation. *Materials Letters* 2013; 97: 67-70.

[968] Ancion PY, Lear G, Neale M et al. Using biofilm as a novel approach to assess stormwater treatment efficacy. *Water Research* 2014; 49: 406-415.

[969] Hu X, Li S, Rong Y et al. Effect of Cu2+ on biofilm and extracellular polymeric substance. *Huagong Xuebao/CIESC Journal* 2014; 65: 1062-1067.

[970] Agarwala M, Choudhury B, Yadav RNS. Comparative Study of Antibiofilm Activity of Copper Oxide and Iron Oxide Nanoparticles Against Multidrug Resistant Biofilm Forming Uropathogens. *Indian Journal of Microbiology* 2014; 54: 365-368.

[971] Fenoglio I, Fubini B, Ghibaudi EM et al. Multiple aspects of the interaction of biomacromolecules with inorganic surfaces. *Advanced Drug Delivery Reviews* 2011; 63: 1186-1209.

[972] Chorianopoulos NG, Tsoukleris DS, Panagou EZ et al. Use of titanium dioxide (TiO2) photocatalysts as alternative means for Listeria monocytogenes biofilm disinfection in food processing. *Food Microbiology* 2011; 28: 164-170.

[973] Nieto Pozo I, Olmos D, Orgaz B et al. Titania nanoparticles prevent development of Pseudomonas fluorescens biofilms on polystyrene surfaces. *Materials Letters* 2014; 127: 1-3.

[974] Dunlop PSM, Sheeran CP, Byrne JA et al. Inactivation of clinically relevant pathogens by photocatalytic coatings. *Journal of Photochemistry and Photobiology A: Chemistry* 2010; 216: 303-310.

[975] Wood P, Caldwell DE, Evans E et al. Surface-catalysed disinfection of thick Pseudomonas aeruginosa biofilms. *Journal of Applied Microbiology* 1998; 84: 1092-1098.

[976] Lequette Y, Boels G, Clarisse M et al. Using enzymes to remove biofilms of bacterial isolates sampled in the food-industry. *Biofouling* 2010; 26: 421-431.

[977] Craigen B, Dashiff A, Kadouri DE. The use of commercially available alpha-amylase compounds to inhibit and remove Staphylococcus aureus biofilms. *Open Microbiology Journal* 2011; 5: 21-31.

[978] Orgaz B, Neufeld RJ, SanJose C. Single-step biofilm removal with delayed release encapsulated Pronase mixed with soluble enzymes. *Enzyme and Microbial Technology* 2007; 40: 1045-1051.

[979] Izano EA, Amarante MA, Kher WB et al. Differential roles of poly-N-acetylglucosamine surface polysaccharide and extracellular DNA in Staphylococcus aureus and Staphylococcus epidermidis biofilms. *Applied and Environmental Microbiology* 2008; 74: 470-476.

[980] Itoh Y, Wang X, Joseph Hinnebusch B et al. Depolymerization of β-1,6-N-acetyl-D-glucosamine disrupts the integrity of diverse bacterial biofilms. *Journal of Bacteriology* 2005; 187: 382-387.

[981] Donelli G, Francolini I, Romoli D et al. Synergistic activity of dispersin B and cefamandole nafate in inhibition of staphylococcal biofilm growth on polyurethanes. *Antimicrobial Agents and Chemotherapy* 2007; 51: 2733-2740.

[982] Darouiche RO, Mansouri MD, Gawande PV et al. Antimicrobial and antibiofilm efficacy of triclosan and DispersinB® combination. *Journal of Antimicrobial Chemotherapy* 2009; 64: 88-93.

[983] Alkawash MA, Soothill JS, Schiller NL. Alginate lyase enhances antibiotic killing of mucoid Pseudomonas aeruginosa in biofilms. *Apmis* 2006; 114: 131-138.

[984] Alipour M, Suntres ZE, Omri A. Importance of DNase and alginate lyase for enhancing free and liposome encapsulated aminoglycoside activity against Pseudomonas aeruginosa. *Journal of Antimicrobial Chemotherapy* 2009; 64: 317-325.

[985] Bayer AS, Park S, Ramos MC et al. Effects of alginase on the natural history and antibiotic therapy of experimental endocarditis caused by mucoid Pseudomonas aeruginosa. *Infection and Immunity* 1992; 60: 3979-3985.

[986] Augustin M, Ali-Vehmas T, Atroshi F. Assessment of enzymatic cleaning agents and disinfectants against bacterial biofilms. *Journal of Pharmacy and Pharmaceutical Sciences* 2004; 7: 55-64.

[987] Batoni G, Maisetta G, Brancatisano FL et al. Use of antimicrobial peptides against microbial biofilms: Advantages and limits. *Current Medicinal Chemistry* 2011; 18: 256-279.

[988] Guerra NP, Araujo AB, Barrera AM et al. Antimicrobial activity of nisin adsorbed to surfaces commonly used in the food industry. *Journal of Food Protection* 2005; 68: 1012-1019.

[989] Tong Z, Zhang L, Ling J et al. An in vitro study on the effect of free amino acids alone or in combination with nisin on biofilms as well as on planktonic bacteria of Streptococcus mutans. *PLoS ONE* 2014; 9.

[990] Malcolm J, Sherriff A, Lappin DF et al. Salivary antimicrobial proteins associate with age-related changes in streptococcal composition in dental plaque. *Molecular Oral Microbiology* 2014;.

[991] Kudva IT, Jelacic S, Tarr PI et al. Biocontrol of Escherichia coli O157 with O157-specific bacteriophages. Applied and Environmental Microbiology *1999; 65: 3767-3773.*

[992] Curtin JJ, Donlan RM. Using bacteriophages to reduce formation of catheter-associated biofilms by Staphylococcus epidermidis. *Antimicrobial Agents and Chemotherapy* 2006; 50: 1268-1275.

[993] Merril CR, Scholl D, Adhya SL. The prospect for bacteriophage therapy in Western medicine. *Nature Reviews Drug Discovery* 2003; 2: 489-497.

[994] Goldman G, Starosvetsky J, Armon R. Inhibition of biofilm formation on UF membrane by use of specific bacteriophages. *Journal of Membrane Science* 2009; 342: 145-152.

[995] Patel J, Sharma M, Millner P et al. Inactivation of Escherichia coli O157:H7 attached to spinach harvester blade using bacteriophage. *Foodborne Pathogens and Disease* 2011; 8: 541-546.

[996] Corbin BD, McLean RJC, Aron GM. Bacteriophage T4 multiplication in a glucose-limited Escherichia coli biofilm. *Canadian Journal of Microbiology* 2001; 47: 680-684.

[997] Sillankorva S, Neubauer P, Azeredo J. Pseudomonas fluorescens biofilms subjected to phage phiIBB-PF7A. *BMC Biotechnology* 2008; 8.

[998] Cerca N, Oliveira R, Azeredo J. Susceptibility of Staphylococcus epidermidis planktonic cells and biofilms to the lytic action of staphylococcus bacteriophage K. *Letters in Applied Microbiology* 2007; 45: 313-317.

[999] Tait K, Skillman LC, Sutherland IW. The efficacy of bacteriophage as a method of biofilm eradication. *Biofouling* 2002; 18: 305-311.
[1000] Lu TK, Collins JJ. Dispersing biofilms with engineered enzymatic bacteriophage. *Proceedings of the National Academy of Sciences of the United States of America* 2007; 104: 11197-11202.
[1001] Persico P, Ambrogi V, Carfagna C et al. Nanocomposite polymer films containing carvacrol for antimicrobial active packaging. *Polymer Engineering and Science* 2009; 49: 1447-1455.
[1002] Zinoviadou KG, Koutsoumanis KP, Biliaderis CG. Physical and thermo-mechanical properties of whey protein isolate films containing antimicrobials, and their effect against spoilage flora of fresh beef. *Food Hydrocolloids* 2010; 24: 49-59.
[1003] Ahmad M, Benjakul S, Prodpran T et al. Physico-mechanical and antimicrobial properties of gelatin film from the skin of unicorn leatherjacket incorporated with essential oils. *Food Hydrocolloids* 2012; 28: 189-199.
[1004] Giménez B, Gómez-Guillén MC, López-Caballero ME et al. Role of sepiolite in the release of active compounds from gelatin-egg white films. *Food Hydrocolloids* 2012; 27: 475-486.
[1005] Türe H, Gällstedt M, Hedenqvist MS. Antimicrobial compression-moulded wheat gluten films containing potassium sorbate. *Food Research International* 2012; 45: 109-115.
[1006] Emiroğlu ZK, Yemiş GP, Coşkun et al. Antimicrobial activity of soy edible films incorporated with thyme and oregano essential oils on fresh ground beef patties. *Meat Science* 2010; 86: 283-288.
[1007] Sivarooban T, Hettiarachchy NS, Johnson MG. Physical and antimicrobial properties of grape seed extract, nisin, and EDTA incorporated soy protein edible films. *Food Research International* 2008; 41: 781-785.
[1008] Oussalah M, Caillet S, Salmiéri S et al. Antimicrobial and antioxidant effects of milk protein-based film containing essential oils for the preservation of whole beef muscle. *Journal of Agricultural and Food Chemistry* 2004; 52: 5598-5605.
[1009] Salgado PR, López-Caballero ME, Gómez-Guillén MC et al. Sunflower protein films incorporated with clove essential oil have potential application for the preservation of fish patties. *Food Hydrocolloids* 2013; 33: 74-84.
[1010] Rojas-Graü MA, Avena-Bustillos RJ, Olsen C et al. Effects of plant essential oils and oil compounds on mechanical, barrier and

antimicrobial properties of alginate-apple puree edible films. *Journal of Food Engineering* 2007; 81: 634-641.

[1011] Leceta I, Guerrero P, Ibarburu I et al. Characterization and antimicrobial analysis of chitosan-based films. *Journal of Food Engineering* 2013; 116: 889-899.

[1012] Li LH, Deng JC, Deng HR et al. Preparation, characterization and antimicrobial activities of chitosan/Ag/ZnO blend films. *Chemical Engineering Journal* 2010; 160: 378-382.

[1013] Pranoto Y, Rakshit SK, Salokhe VM. Enhancing antimicrobial activity of chitosan films by incorporating garlic oil, potassium sorbate and nisin. *LWT - Food Science and Technology 2005*; 38: 859-865.

[1014] Rao MS, Kanatt SR, Chawla SP et al. Chitosan and guar gum composite films: Preparation, physical, mechanical and antimicrobial properties. *Carbohydrate Polymers* 2010; 82: 1243-1247.

[1015] Ramos TL, Silva SI, Soares JC et al. Features and performance of edible films, obtained from whey protein isolate formulated with antimicrobial compounds. *Food Research International* 2012; 45: 351-361.

[1016] Sayanjali S, Ghanbarzadeh B, Ghiassifar S. Evaluation of antimicrobial and physical properties of edible film based on carboxymethyl cellulose containing potassium sorbate on some mycotoxigenic Aspergillus species in fresh pistachios. *LWT - Food Science and Technology* 2011; 44: 1133-1138.

[1017] Mani-López E, García HS, López-Malo A. Organic acids as antimicrobials to control Salmonella in meat and poultry products. *Food Research International* 2012; 45: 713-721.

[1018] Campos CA, Gerschenson LN, Flores SK. Development of Edible Films and Coatings with Antimicrobial Activity. *Food and Bioprocess Technology* 2011; 4: 849-875.

[1019] Boulange-Petermann L, Jullien C, Dubois PE et al. Influence of surface chemistry on the hygienic status of industrial stainless steel. *Biofouling* 2004; 20: 25-33.

[1020] Wu ZH, Chen HB, Dong YM et al. Cleaning using nanobubbles: Defouling by electrochemical generation of bubbles. *Journal of Colloid and Interface Science* 2008; 328: 10-14.

[1021] Cornelissen ER, Vrouwenvelder JS, Heijman SGJ et al. Periodic air/water cleaning for control of biofouling in spiral wound membrane elements. *Journal of Membrane Science* 2007; 287: 94-101.

[1022] Psoch C, Schiewer S. Direct filtration of natural and simulated river water withair sparging and sponge ball application for fouling control. *Desalination* 2006; 197: 190-204.

[1023] Cloete TE. Biofouling control in industrial water systems: What we know and what we need to know. *Materials and Corrosion* 2003; 54: 520-526.

[1024] Holah JT, Thorpe RH. Cleanability in relation to bacterial retention on unused and abraded domestic sink materials. *Journal of Applied Bacteriology* 1990; 69: 599-608.

[1025] Parkar SG, Flint SH, Brooks JD. Evaluation of the effect of cleaning regimes on biofilms of thermophilic bacilli on stainless steel. *Journal of Applied Microbiology* 2004; 96: 110-116.

[1026] Antoniou K, Frank JF. Removal of Pseudomonas putida biofilm and associated extracellular polymeric substances from stainless steel by alkali cleaning. *Journal of Food Protection* 2005; 68: 277-281.

[1027] Srinivasan R, Stewart PS, Griebe T et al. Biofilm parameters influencing biocide efficacy. *Biotechnology and Bioengineering* 1995; 46: 553-560.

[1028] Bremer PJ, Monk I, Butler R. Inactivation of Listeria monocytogenes/Flavobacterium spp. biofilms using chlorine: Impact of substrate, pH, time and concentration. Letters *in Applied Microbiology* 2002; 35: 321-325.

[1029] Kuda T, Yano T, Kuda MT. Resistances to benzalkonium chloride of bacteria dried with food elements on stainless steel surface. *LWT - Food Science and Technology* 2008; 41: 988-993.

[1030] Sadoudi AK, Herry JM, Cerf O. Elimination of adhering bacteria from surfaces by pulsed laser beams. *Letters in Applied Microbiology* 1997; 24: 177-179.

[1031] Ryther R. Food Technologies: Cleaning and Disinfection Technologies (Clean-In-Place, Clean-Out-of-Place). In: Motarjemi Y, Hrsg. *Encyclopedia of Food Safety.* Waltham: Academic Press; 2014:149-155.

[1032] Berk Z. Chapter 28 - Cleaning, Disinfection, Sanitation. In: Berk Z, Hrsg. *Food Process Engineering and Technology* (Second Edition). San Diego: Academic Press; 2013:637-650.

[1033] Forsythe SJ, Hayes PR. *Food Hygiene,* Microbiology and HACCP 1998;.

[1034] Bremer PJ, Fillery S, McQuillan AJ. Laboratory scale Clean-In-Place (CIP) studies on the effectiveness of different caustic and acid wash steps on the removal of dairy biofilms. *International Journal of Food Microbiology* 2006; 106: 254-262.

[1035] Kumar CG, Anand SK. Significance of microbial biofilms in food industry: A review. *International Journal of Food Microbiology 1998*; 42: 9-27.

[1036] Dufour M, Simmonds RS, Bremer PJ. Development of a laboratory scale clean-in-place system to test the effectiveness of "natural" antimicrobials against dairy biofilms. *Journal of Food Protection* 2004; 67: 1438-1443.

[1037] Sharma M, Anand SK. Characterization of constitutive microflora of biofilms in dairy processing lines. *Food Microbiology* 2002; 19: 627-636.

[1038] Eide MH, Homleid JP, Mattsson B. Life cycle assessment (LCA) of cleaning-in-place processes in dairies. *LWT - Food Science and Technology* 2003; 36: 303-314.

[1039] Lelièvre C, Antonini G, Faille C et al. Cleaning-in-place: Modelling of cleaning kinetics of pipes soiled by Bacillus spores assuming a process combining removal and deposition. *Food and Bioproducts Processing: Transactions of the Institution of of Chemical Engineers,* Part C 2002; 80: 305 311.

[1040] Niemira BA. Irradiation sensitivity of planktonic and biofilm-associated Listeria monocytogenes and L. innocua as influenced by temperature of biofilm formation. *Food and Bioprocess Technology* 2010; 3: 257-264.

[1041] Khaw A, Logan R, Keefe D et al. Radiation-induced oral mucositis and periodontitis - proposal for an inter-relationship. *Oral Diseases* 2014; 20: e7-e18.

[1042] Feuerstein O. Light therapy: complementary antibacterial treatment of oral biofilm. *Advances in dental research* 2012; 24: 103-107.

[1043] Bernbom N, Vogel BF, Gram L. Listeria monocytogenes survival of UV-C radiation is enhanced by presence of sodium chloride, organic food material and by bacterial biofilm formation. *International Journal of Food Microbiology* 2011; 147: 69-73.

[1044] Vleugels M, Shama G, Deng XT et al. Atmospheric plasma inactivation of biofilm-forming bacteria for food safety control. *31st IEEE International Conference on Plasma Science* (ICOPS 2004) 2004: 824-828.

[1045] Oulahal-Lagsir N, Martial-Gros A, Bonneau M et al. "Escherichia coli-milk" biofilm removal from stainless steel surfaces: Synergism between ultrasonic waves and enzymes. *Biofouling 2003*; 19: 159-168.

[1046] Pezzoni M, Pizarro RA, Costa CS. Protective role of extracellular catalase (KatA) against UVA radiation in Pseudomonas aeruginosa biofilms. *Journal of Photochemistry and Photobiology B: Biology* 2014; 131: 53-64.

[1047] Bott TR, Tianqing L. Ultrasound enhancement of biocide efficiency. *Ultrasonics* Sonochemistry 2004; 11: 323-326.

[1048] Baumann AR, Martin SE, Feng H. Removal of Listeria monocytogenes biofilms from stainless steel by use of ultrasound and Ozone. *Journal of Food Protection* 2009; 72: 1306-1309.

[1049] Oulahal-Lagsir N, Martial-Gros A, Bonneau M et al. Ultrasonic methodology coupled to ATP bioluminescence for the non-invasive detection of fouling in food processing equipment - Validation and application to a dairy factory. *Journal of Applied Microbiology* 2000; 89: 433-441.

[1050] Erriu M, Blus C, Szmukler-Moncler S et al. Microbial biofilm modulation by ultrasound: Current concepts and controversies. *Ultrasonics Sonochemistry* 2014; 21: 15-22.

[1051] Kirzhner F, Zimmels Y, Malkovskaja A et al. Removal of microbial biofilm on Water Hyacinth plants roots by ultrasonic treatment. *Ultrasonics* 2009; 49: 153-158.

[1052] Nithila SDR, Anandkumar B, Vanithakumari SC et al. Studies to control biofilm formation by coupling ultrasonication of natural waters and anodization of titanium. *Ultrasonics Sonochemistry* 2014; 21: 189-199.

[1053] Iqbal K, Ohl S-W, Khoo B-C et al. Effect of High-Intensity Focused Ultrasound on Enterococcus Faecalis Planktonic Suspensions and Biofilms. *Ultrasound in Medicine & Biology* 2013; 39: 825-833.

[1054] Piyasena P, Mohareb E, McKellar RC. Inactivation of microbes using ultrasound: A review. *International Journal of Food Microbiology* 2003; 87: 207-216.

[1055] Torlak E, Sert D. Combined effect of benzalkonium chloride and ultrasound against Listeria monocytogenes biofilm on plastic surface. *Letters in Applied Microbiology* 2013; 57: 220-226.

[1056] Pitt WG, Ross SA. Ultrasound increases the rate of bacterial cell growth. *Biotechnology Progress* 2003; 19: 1038-1044.

[1057] Peterson RV, Pitt WG. The effect of frequency and power density on the ultrasonically-enhanced killing of biofilm-sequestered Escherichia coli. *Colloids and Surfaces B: Biointerfaces* 2000; 17: 219-227.

[1058] Carmen JC, Nelson JL, Beckstead BL et al. Ultrasonic-enhanced gentamicin transport through colony biofilms of Pseudomonas aeruginosa and Escherichia coli. *Journal of Infection and Chemotherapy 2004*; 10: 193-199.

[1059] Carmen JC, Roeder BL, Nelson JL et al. Treatment of biofilm infections on implants with low-frequency ultrasound and antibiotics. *American Journal of Infection Control* 2005; 33: 78-82.

[1060] Thurnheer T, Rohrer E, Belibasakis GN et al. Static biofilm removal around ultrasonic tips in vitro. *Clinical Oral Investigations 2013*: 1-6.

[1061] Agarwal A, Jern Ng W, Liu Y. Removal of biofilms by intermittent low-intensity ultrasonication triggered bursting of microbubbles. *Biofouling* 2014; 30: 359-365.

[1062] Cos P, Toté K, Horemans T et al. Biofilms: An extra hurdle for effective antimicrobial therapy. *Current Pharmaceutical Design* 2010; 16: 2279-2295.

[1063] Ha JH, Ha SD. Synergistic effects of sodium hypochlorite and ultraviolet radiation in reducing the levels of selected foodborne pathogenic bacteria. *Foodborne Pathogens and Disease* 2011; 8: 587-591.

[1064] DeQueiroz GA, Day DF. Antimicrobial activity and effectiveness of a combination of sodium hypochlorite and hydrogen peroxide in killing and removing Pseudomonas aeruginosa biofilms from surfaces. *Journal of Applied Microbiology* 2007; 103: 794-802.

[1065] Vankerckhoven E, Verbessem B, Crauwels S et al. Exploring the potential synergistic effects of chemical disinfectants and UV on the inactivation of free-living bacteria and treatment of biofilms in a pilot-scale system. *Water Science and Technology* 2011; 64: 1247-1253.

[1066] Schurman JJ. Antibacterial Activity of Hydrogen Peroxide Against Escherichia ColiO157:H7 and Salmonella Spp in Fruit Juices, *Both Alone and in Combination with Organic Acids* 2001;.

[1067] Patil S. Efficacy of ozone and ultrasound for microbial reduction in fruit juice 2010;.

[1068] Verma V, Harjai K, Chhibber S. Structural changes induced by a lytic bacteriophage make ciprofloxacin effective against older biofilm of Klebsiella pneumoniae. *Biofouling* 2010; 26: 729-737.

[1069] Haaken D, Dittmar T, Schmalz V et al. Disinfection of biologically treated wastewater and prevention of biofouling by UV/electrolysis hybrid technology: Influence factors and limits for domestic wastewater reuse. *Water Research* 2014; 52: 20-28.

[1070] Kim JG, Yousef AE, Khadre MA. Ozone and its current and future application in the food industry. In, *Advances in Food and Nutrition Research;* 2003:167-218.

[1071] Johansen C, Falholt P, Gram L. Enzymatic removal and disinfection of bacterial biofilms. *Applied and Environmental Microbiology* 1997; 63: 3724-3728.

[1072] Russell SM. The effect of electrolyzed oxidative water applied using electrostatic spraying on pathogenic and indicator bacteria on the surface of eggs. *Poultry Science* 2003; 82: 158-162.

[1073] Russell AD. Mechanisms of action, resistance, and stress adaptation. *Antimicrobials in food* 2005: 633-657.

[1074] Anwar H, Strap JL, Costerton JW. Susceptibility of biofilm cells of Pseudomonas aeruginosa to bactericidal actions of whole blood and serum. *FEMS Microbiology Letters* 1992; 92: 235-241.

[1075] Rutala WA, Weber DJ. The Healthcare Infection Control Practices Advisory Committee. *CDC guideline for disinfection and sterilization in healthcare facilities* 0000;.

[1076] Niemira BA. Relative efficacy of sodium hypochlorite wash versus irradiation to inactivate Escherichia coli O157:H7 internalized in leaves of romaine lettuce and baby spinach. *Journal of Food Protection* 2007; 70: 2526-2532.

[1077] Brandl MT, Mandrell RE. Fitness of Salmonella enterica serovar Thompson in the cilantro phyllosphere. *Applied and Environmental Microbiology* 2002; 68: 3614-3621.

[1078] Hassan AN, Frank JF. Attachment of Escherichia coli O157:H7 grown in tryptic soy broth and nutrient broth to apple and lettuce surfaces as related to cell hydrophobicity, surface charge, and capsule production. *International Journal of Food Microbiology* 2004; 96: 103-109.

[1079] Stewart D, Reineke K, Ulaszek J et al. Growth of Escherichia coli O157:H7 during sprouting of alfalfa seeds. *Letters in Applied Microbiology* 2001; 33: 95-99.

[1080] Warriner K, Spaniolas S, Dickinson M et al. Internalization of bioluminescent Escherichia coli and Salmonella Montevideo in growing bean sprouts. *Journal of Applied Microbiology* 2003; 95: 719-727.

[1081] Behnke S, Parker AE, Woodall D et al. Comparing the chlorine disinfection of detached biofilm clusters with those of sessile biofilms and planktonic cells in single- and dual-species cultures. *Applied and Environmental Microbiology* 2011; 77: 7176-7184.

[1082] Weiss A, Hammes WP. Efficacy of heat treatment in the reduction of salmonellae and Escherichia coli O157:H- on alfalfa, mung bean and radish seeds used for sprout production. *European Food Research and Technology* 2005; 221: 187-191.

[1083] Luppens SBI, Reij MW, Van der Heijden RWL et al. Development of a standard test to assess the resistance of Staphylococcus aureus biofilm cells to disinfectants. *Applied and Environmental Microbiology* 2002; 68: 4194-4200.

[1084] Wirtanen G, Mattila-Sandholm T. Effect of the growth phase of foodborne biofilms on their resistance to a chlorine sanitizer. Part II. *LWT - Food Science and Technology* 1992; 25: 50-54.

[1085] Sommer P, Martin-Rouas C, Mettler E. Influence of the adherent population level on biofilm population, structure and resistance to chlorination. *Food Microbiology* 1999; 16: 503-515.

[1086] Mangalappalli-Illathu AK, Vidović S, Korber DR. Differential adaptive response and survival of Salmonella enterica serovar enteritidis planktonic and biofilm cells exposed to benzalkonium chloride. *Antimicrobial Agents and Chemotherapy* 2008; 52: 3669-3680.

[1087] Percival SL, Walker JT, Hunter PR. *Microbiological Aspects of Biofilms and Drinking Water* 2000;.

[1088] Manuzon MY, Wang HH. Mixed Culture Biofilms. *Biofilms in the Food Environment* 2007: 105-126.

[1089] Gilbert P, Allison DG, McBain AJ. Biofilms in vitro and in vivo: Do singular mechanisms imply cross-resistance? *Journal of Applied Microbiology Symposium Supplement* 2002; 92: 98S-110S.

[1090] Stewart PS, Mukherjee PK, Ghannoum MA. Biofilm antimicrobial resistance. *Microbial Biofilms* 2004; 1: 250-268.

[1091] Del Pozo JL, Patel R. The challenge of treating biofilm-associated bacterial infections. *Clinical Pharmacology and Therapeutics* 2007; 82: 204-209.

[1092] Leriche V, Briandet R, Carpentier B. Ecology of mixed biofilms subjected daily to a chlorinated alkaline solution: Spatial distribution of bacterial species suggests a protective effect of one species to another. *Environmental Microbiology* 2003; 5: 64-71.

[1093] Chorianopoulos NG, Giaouris ED, Skandamis PN et al. Disinfectant test against monoculture and mixed-culture biofilms composed of technological, spoilage and pathogenic bacteria: Bactericidal effect of essential oil and hydrosol of Satureja thymbra and comparison with standard acid-base sanitizers. *Journal of Applied Microbiology* 2008; 104: 1586-1596.

[1094] Frank JF, Koffi RA. Surface-adherent growth of Listeria monocytogenes is associated with increased resistance to surfactant sanitizers and heat. *J Food Prot* 1990; 53: 550-554.

[1095] Mafu AA, Roy D, Goulet J et al. Efficiency of sanitizing agents for destroying Listeria monocytogenes on contaminated surfaces. *Journal of dairy science* 1990; 73: 3428-3432.

[1096] Suci PA, Mittelman MW, Yu FP et al. Investigation of ciprofloxacin penetration into Pseudomonas aeruginosa biofilms. *Antimicrobial Agents and Chemotherapy* 1994; 38: 2125-2133.

[1097] Brown MR, Gilbert P. Sensitivity of biofilms to antimicrobial agents. *Journal of Applied* Bacteriology 1993; 74 Suppl: 87S-97S.

[1098] Gilbert P, Collier PJ, Brown MRW. Influence of growth rate on susceptibility to antimicrobial agents: Biofilms, cell cycle, dormancy, and strigent response. *Antimicrobial Agents and Chemotherapy* 1990; 34: 1865-1868.

[1099] Evans DJ, Allison DG, Brown MRW et al. Susceptibility of Pseudomonas aeruginosa and Escherichia coli biofilms towards ciprofloxacin: effect of specific growth rate. *Journal of Antimicrobial Chemotherapy* 1991; 27: 177-184.

[1100] Lee SH, Frank JF. Inactivation of surface-adherent Listeria monocytogenes hypochlorite and heat. *J Food Prot* 1991; 54: 4-6.

[1101] Humphrey TJ, Slater E, McAlpine K et al. Salmonella enteritidis phage type 4 isolates more tolerant of heat, acid, or hydrogen peroxide also survive longer on surfaces. *Applied and Environmental Microbiology* 1995; 61: 3161-3164.

[1102] Szomolay B, Klapper I, Dockery J et al. Adaptive responses to antimicrobial agents in biofilms. *Environmental Microbiology* 2005; 7: 1186-1191.

[1103] Sanderson SS, Stewart PS. Evidence of bacterial adaptation to monochloramine in Pseudomonas aeruginosa biofilms and evaluation of biocide action model. *Biotechnology and Bioengineering* 1997; 56: 201-209.

[1104] Amorena B, Gracia E, Monzón M et al. Antibiotic susceptibility assay for Staphylococcus aureus in biofilms developed in vitro. *Journal of Antimicrobial Chemotherapy* 1999; 44: 43-55.

[1105] Hoffman LR, D'Argenio DA, MacCoss MJ et al. Aminoglycoside antibiotics induce bacterial biofilm formation. *Nature* 2005; 436: 1171-1175.

[1106] Molin S, Tolker-Nielsen T. Gene transfer occurs with enhanced efficiency in biofilms and induces enhanced stabilisation of the biofilm structure. *Current Opinion in Biotechnology* 2003; 14: 255-261.

[1107] Percival SL, Hill KE, Malic S et al. Antimicrobial tolerance and the significance of persister cells in recalcitrant chronic wound biofilms. *Wound Repair and Regeneration* 2011; 19: 1-9.

[1108] Lewis K. Persister cells, dormancy and infectious disease. *Nature Reviews Microbiology* 2007; 5: 48-56.

[1109] Kolodkin-Gal I, Romero D, Cao S et al. D-Amino acids trigger biofilm disassembly. *Science* 2010; 328: 627-629.

[1110] Kolodkin-Gal I, Cao S, Chai L et al. A self-produced trigger for biofilm disassembly that targets exopolysaccharide. *Cell* 2012; 149: 684-692.

[1111] Pavlukhina SV, Kaplan JB, Xu L et al. Noneluting enzymatic antibiofilm coatings. *ACS Applied Materials and Interfaces* 2012; 4: 4708-4716.

[1112] Jennings JA, Courtney HS, Haggard WO. Cis-2-decenoic acid inhibits S. aureus growth and biofilm in vitro: A pilot study basic research. *Clinical Orthopaedics and Related Research* 2012; 470: 2663-2670.

[1113] Barraud N, Storey MV, Moore ZP et al. Nitric oxide-mediated dispersal in single- and multi-species biofilms of clinically and industrially relevant microorganisms. *Microbial Biotechnology* 2009; 2: 370-378.

[1114] Torres-Giner S. *Adv Biomater* 2012; 14: 112-B122.

[1115] Trautner BW, Hull RA, Thornby JI et al. Coating urinary catheters with an avirulent strain of Escherichia coli as a means to establish asymptomatic colonization. *Infection Control and Hospital Epidemiology* 2007; 28: 92-94.

[1116] Ishijima SA, Hayama K, Burton JP et al. Effect of Streptococcus salivarius K12 on the In vitro growth of Candida albicans and its protective effect in an oral candidiasis model. *Applied and Environmental Microbiology* 2012; 78: 2190-2199.

[1117] Catuogno C, Jones MN. The antibacterial properties of solid supported liposomes on Streptococcus oralis biofilms. *International Journal of Pharmaceutics* 2003; 257: 125-140.

[1118] Fett WF. Naturally occurring biofilms on alfalfa and other types of sprouts. *Journal of Food Protection 2000*; 63: 625-632.

[1119] Hangler M, Burmølle M, Schneider I et al. The serine protease Esperase HPF inhibits the formation of multispecies biofilm. *Biofouling* 2009; 25: 667-674.

[1120] Schwering M, Song J, Louie M et al. Multi-species biofilms defined from drinking water microorganisms provide increased protection against chlorine disinfection. *Biofouling* 2013; 29: 917-928.

[1121] Shu M, Browngardt CM, Chen YYM et al. Role of Urease Enzymes in Stability of a 10-Species Oral Biofilm Consortium Cultivated in a Constant-Depth Film Fermenter. *Infection and Immunity* 2003; 71: 7188-7192.

[1122] Lee KW. *Biofilm development and enhanced stress resistance of a model, mixed-species community biofilm.* Isme j 2013;.

[1123] Shirtliff ME, Peters BM, Jabra-Rizk MA. Cross-kingdom interactions: Candida albicans and bacteria. *FEMS Microbiology Letters* 2009; 299: 1-8.

[1124] Harriott MM, Noverr MC. Candida albicans and Staphylococcus aureus form polymicrobial biofilms: Effects on antimicrobial resistance. *Antimicrobial Agents and Chemotherapy* 2009; 53: 3914-3922.

[1125] Peters BM, Jabra-Rizk MA, Scheper MA et al. Microbial interactions and differential protein expression in Staphylococcus aureus -Candida albicans dual-species biofilms. *FEMS Immunology and Medical Microbiology* 2010; 59: 493-503.

[1126] Langsrud S, Moretro T, Heir E. Bacteria in the food production environment. *Food Science and Technology (London)* 2012; 26: 26-30.

[1127] Verran J. Biofouling in food processing: Biofilm or biotransfer potential? *Food and Bioproducts Processing: Transactions of the Institution of of Chemical Engineers, Part C* 2002; 80: 292-298.

[1128] Bjarnsholt T, Kirketerp-Møller K, Jensen PØ et al. Why chronic wounds will not heal: A novel hypothesis. *Wound Repair and Regeneration 2008*; 16: 2-10.

INDEX

B

C

E

F

G

H

I

J

K

L

M

N

O

P

Q

R

S

T